Practical Bioinformatics

AF801476

Sufang Wang • Michael Gribskov

Practical Bioinformatics

A Laboratory Manual

Springer

Sufang Wang
School of Life Sciences
Northwestern Polytechnical University
Xi'an, Shaanxi, China

Michael Gribskov
Department of Biological Sciences
Purdue University
West Lafayette, IN, USA

ISBN 978-981-96-7948-5 ISBN 978-981-96-7949-2 (eBook)
https://doi.org/10.1007/978-981-96-7949-2

© The Editor(s) (if applicable) and The Author(s), under exclusive license to Springer Nature Singapore Pte Ltd. 2025

This work is subject to copyright. All rights are solely and exclusively licensed by the Publisher, whether the whole or part of the material is concerned, specifically the rights of translation, reprinting, reuse of illustrations, recitation, broadcasting, reproduction on microfilms or in any other physical way, and transmission or information storage and retrieval, electronic adaptation, computer software, or by similar or dissimilar methodology now known or hereafter developed.
The use of general descriptive names, registered names, trademarks, service marks, etc. in this publication does not imply, even in the absence of a specific statement, that such names are exempt from the relevant protective laws and regulations and therefore free for general use.
The publisher, the authors and the editors are safe to assume that the advice and information in this book are believed to be true and accurate at the date of publication. Neither the publisher nor the authors or the editors give a warranty, expressed or implied, with respect to the material contained herein or for any errors or omissions that may have been made. The publisher remains neutral with regard to jurisdictional claims in published maps and institutional affiliations.

This Springer imprint is published by the registered company Springer Nature Singapore Pte Ltd.
The registered company address is: 152 Beach Road, #21-01/04 Gateway East, Singapore 189721, Singapore

If disposing of this product, please recycle the paper.

Preface

The field of bioinformatics is advancing at a remarkable rate. With the rapid development of new analytical techniques that use the latest advances in machine learning and data science, today's biologists are gaining fantastic new insights into the natural world's most complex systems. An increasing number of students are interested in learning bioinformatics, but most of the bioinformatics textbooks are theory-based and lack a lab-based manual that can assist them in handling large biological data. Therefore, it is necessary and urgent to have a lab manual that can be integrated with bioinformatics courses.

This book is aimed at students, faculty, and researchers in biology, health/medical sciences, agricultural sciences, bioengineering, biotechnology, and genetic engineering. The content should be easily accessible to undergraduate students, faculty, and general readers interested in biology or bioinformatics.

As a motivating example, the first ten chapters focus on bioinformatics approaches to the study of antibiotic resistance. No previous experience in bioinformatics is assumed. The last two chapters focus on the use of R to analyze and solve biological questions. Users need to install R or R studio and related packages on a local computer. Again, no previous experience in R/R studio is needed. All examples and programming codes in this book are designed for beginners. These 12 chapters can be used as the basis for a 6- to 12-week laboratory module. This book can be used either stand-alone, or in conjunction with a lecture course.

Resources and Materials

This book uses several online databases, compute servers, and freely available software. Each lab lists the sites/software needed for the exercise, but to make preparation easier, here is a complete list:

- NCBI
  - Entrez
  - Blast
  - CDD
  - Cn3D
- EBI
  - Software dispatcher
  - Interpro
- UniProt
- GTDB
- PDB
- R
- R Studio
- Bioconductor
  - DESeq2
  - Pheatmap

Laboratory Notebook

Because this is a laboratory manual, we assume that students will keep their results in a laboratory notebook. Clearly an electronic notebook will be easiest for the analyses and results of the activities we present here. However, the possible options are legion and will depend on the educational technology solutions in place at different institutions. The critical features are that the notebook software should allow students to answer the questions in each exercise, include their numeric and graphical results, and most important of all, provide a mechanism for instructor feedback. Our recommendation is to use Microsoft Word because the students are generally familiar with it, and it is easy to cut and paste results from web sites, to do simple calculations using Excel, and to make annotated figures using PowerPoint. Another option I have successfully used is a Wiki; students are often unfamiliar with wikis, but they are simple to learn and provide the necessary features. There are many electronic laboratory notebooks available—many of them free for academic use. I have had limited experience with *Benchling*, which is one option you could consider. Many universities in the United States use the *Brightspace* course management system. In principle this system could be used, but it seems that it might be quite cumbersome so I haven't actually tried.

Each laboratory includes questions to be answered before the lab. The main purpose of the *Before Lab Questions* is to ensure that the students have read the introductory material before the lab begins. These questions can be included in the lab notebook, or turned in separately. In the activities, results to be included in the notebook and questions to be answered and included are shown in a Text Box, like the following:

> **Notebook**
> Include this plot in your notebook. Describe why this result confirms that the sequences are homologous.

Xi'an, China
West Lafayette, IN, USA

Sufang Wang
Michael Gribskov

Competing Interests

The authors have no competing interests to declare that are relevant to the content of this manuscript.

Contents

Chapter 1
Introduction to Bioinformatics

1.1 What Is Bioinformatics?

The three greatest projects of the twentieth century are the Manhattan project (1942–1945), the Apollo project (1961–1972), and human genome project (1990–2003). Each one had its own merits. The human genome project was an international scientific research project, which aimed to determine human DNA sequence and gene function. It was formally launched in 1990 and completed in April 2003. After the human genome was completed, many other organisms such as Arabidopsis, mouse, and zebra fish have been sequenced and analyzed. In July 2025, the US National Center for Biotechnology Information contained complete genomic sequences of 53,232 eukaryotes (animals, fungi, and plants), 2,692,439 prokaryotes (bacteria), 230,411 viruses, a well as many sequences of organellar genome (chloroplast and mitochondiron), plasmids, and individual genes.

Until the 1980s, bioinformatics was not a widely used term. The most common terms were biocomputing or computational biology. Bioinformatics appears to have been coined by Paulien Hogeweg and Ben Hesper in the early 1970s,[1] who defined it as "the study of informatic processes in biotic systems." Many of the earliest uses, however, were exclusively in Dutch, so the term did not become common in English until later. My personal recollection is of hearing it at the "Macromolecules, Genes and Computers" workshop in 1986. There was considerable resistance to the label because many people felt bioinformatics was restricted to topics related to information storage and retrieval, although this was clearly not the intent of Hogeweg and Hesper. In any case, in spite of the difficulty of referring to oneself as a bioinformaticist, which doesn't easily roll off the tongue, it did become the dominant name for the field. Along the way, the definition reverted to something much closer to the original. Currently, almost any computational analysis of macromolecules

[1]Hogeweg P. The roots of bioinformatics in theoretical biology. PLoS Comput Biol. 2011: e1002021. doi: 10.1371/journal.pcbi.1002021.

© The Author(s), under exclusive license to Springer Nature Singapore Pte Ltd. 2025

S. Wang, M. Gribskov, *Practical Bioinformatics*,
https://doi.org/10.1007/978-981-96-7949-2_1

and/or biological information flow can be considered as bioinformatics, including topics such as genome assembly and annotation, protein structure prediction and analysis, biological network analysis, phylogenetic tree construction, high-throughput sequencing data analysis, biological image analysis, and all of the various "omics."

1.2 Sequence Technology

Bioinformatics has been driven by the development of sequencing technology, which allows scientists to decipher the secrets of DNA sequences. The first-generation of sequence technology was invented by two different groups. One was a selective cleavage method developed by Allan Maxam and Walter Gilbert (usually referred to as Maxam-Gilbert sequencing), the other was a sequencing by synthesis method developed by Frederick Sanger and coworkers (referred to as chain termination, dideoxy, or Sanger sequencing). Maxam-Gilbert sequencing is no longer in use, while modifications of Sanger's sequencing by the synthesis approach remain widely used. First-generation sequencing was a manual process that involved hand-loading samples onto large polyacrylamide gels. First generation methods were eventually automated with separation of DNA fragments using gel or capillary electrophoresis. Sequencers of this generation, such as the Applied Biosystems PRISM, was the workhorses in producing the human genome (2001).

Second-generation sequencing marked the first massively parallel approach, and was the first approach that did not require amplifying and purifying individual sequence fragments. In pyrosequencing (also known as 454 sequencing), DNA was randomly attached to beads using "adapter" sequences and amplified in situ. The beads were passed over a flowcell with wells that could hold just one bead. The sequencing by synthesis process then proceeded by passing one nucleotide at a time through the flowcell and measuring the amount of pyrophosphate released. However, automated dideoxy sequencing (ABI) and 454 pyrosequencing, were soon replaced by next-generation sequencing (NGS).

Currently, next generation sequencing, NGS, mostly refers to Illumina sequencing. NGS is massively parallel, sequencing millions of fragments (usually the length is 100–300 bp) simultaneously per run. This process translates into sequencing hundreds to thousands of genes at one time. With NGS methods, whole genomes can be rapidly sequenced. For instance, large plant and animal genomes can be completed in days. NGS methods can also be used to sequence messenger RNA (technically a cDNA copy of a RNA). The technique, known as RNA-Seq, can discover novel RNA variants and splice sites or quantify mRNAs for gene expression analysis. NGS is the main cause of the explosion in the number of organisms whose genomes and transcriptomes have been sequenced and analyzed. This technology is currently the most successful and widely used, but long-read sequencing is seeing increasing usage.

Table 1.1 Information obtained from sequencing

Target	Information
DNA	Genome/genomics
Messenger RNA	Transcriptome/transcriptomics
Ribosome RNA-protected fragments	Translatome/translatomics
Protein	Proteome/proteomics
Metabolites	Metabolism/metabolomics
DNA methylation	Epigenome/epigenomics

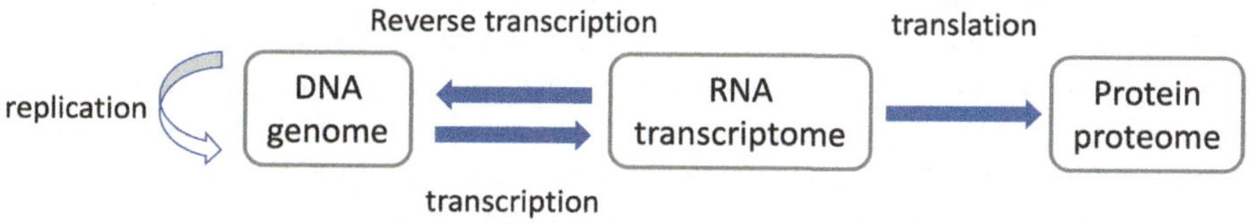

Fig. 1.1 Central dogma and -omics

Single-molecule sequencing (SMS), or long-read sequencing, produces relatively long reads (2000–30,000 bases), although the technology is rapidly advancing at the time this book is being written. Pacific Biosciences and Oxford Nanopore currently produce the most widely used instruments, but there are new companies entering the market.

With the development of sequencing technology, scientists can directly or indirectly sequence many macromolecules, including DNA, RNA, and protein (Table 1.1).

Mass spectrophotometric methods also allow the identification of small molecules (metabolites) as well as directly determining the expressed proteome and sequences of individual proteins. Although many have deplored the usage, the information we obtain is commonly referred to as -omics data: genomics, transcriptomics, proteomics, and many others. In the era of big data, analyses of these data are critically important, and are the focus of this book. DNA, RNA, and proteins are the raw materials of bioinformatics. We expect the students have some prior knowledge of these molecules and of central dogma of biology (Fig. 1.1).

1.3 Topics in Bioinformatics

As mentioned before, bioinformatics today covers almost any analysis of biological systems that includes computational analysis, prediction, or modeling. This includes a very broad range of topics: gene-environment interaction, gene-disease interaction, medical imaging, molecular structure analysis, evolutionary studies, systems biology, and, of course, the development of computational analysis tools and databases. Such a wide range of topics is impossible to cover in a single text. In this manual, we limit ourselves mostly to analyses that are based on primary sequence information (although we do go a bit further in looking at differential gene expression). Most of the analyses in this manual can be performed in time periods of 2–4 h, which is typical for a life science laboratory course. All the analyses can be performed using freely available software, and in most cases, using Internet accessible servers. We hope this will make the material widely available to students at all levels.

Overall, bioinformatics is one of the most fundamental disciplines in the twenty-first century, it integrates knowledge of biology, computer science, mathematics, statistics, physics, and chemistry. In this book, we will introduce the basic concepts in bioinformatics and how to analyze sequence data, which will be beneficial to students in many life science fields. The main topics we cover are centered on what can be learned by identifying homologous genes/proteins. This includes coverage of the concept of homology, sequence database searches, multiple sequence alignments, phylogenetic trees, and functional annotation based on sequence motifs. In addition, we include brief coverage of differential gene expression and machine learning methods which are often used in RNA-seq analysis.

Chapter 2
Introduction to Antibiotic Resistance

Antibiotic resistance, also known as antimicrobial resistance (AMR), is one of the most serious medical problems facing the world today. After a century in which death from bacterial infection was relatively rare, we now face the possibility of returning to the time of our great-grandparents, in which 50% or more of deaths could be due to bacterial infection. Bioinformatics plays a huge role in studying antibiotic resistance. It can be used to identify antibiotic resistance genes, the origin of resistance genes, the mechanism of antibiotic resistance, and, sometimes, possible alternative treatments. This chapter introduces the concept and mechanisms of antibiotic resistance, which we use as an example of how bioinformatics analyses can be used to solve biological questions.

2.1 Objectives

- Highlighting the importance of bacterial resistance to antibiotics.
- Introducing the causes and mechanisms of resistance.

2.2 Importance of Antibiotic Resistance

Bacterial drug resistance, also called antimicrobial resistance (AMR), is a serious worldwide problem. A World Health Organization (WHO) report (April 2014) stated,

> … this serious threat is no longer a prediction for the future, it is happening right now in every region of the world and has the potential to affect anyone, of any age, in any country. Antibiotic resistance—when bacteria change so antibiotics no longer work in people who need them to treat infections—is now a major threat to public health.

© The Author(s), under exclusive license to Springer Nature Singapore Pte Ltd. 2025

S. Wang, M. Gribskov, *Practical Bioinformatics*,
https://doi.org/10.1007/978-981-96-7949-2_2

Table 2.1 Dangerous antibiotic-resistant bacterial strains (2024)

Urgent threats	Serious threats	Concerning	Watch list
Carbapenem-resistant *Acinetobacter*	Drug-resistant *campylobacter*	Erythromycin-resistant group A *streptococcus*	Azole-resistant *aspergillus fumigatus*
Candida auris	Fluconazole-resistant *Candida*	Clindamycin-resistant group B *streptococcus*	Drug-resistant *mycoplasma genitalium*
Clostridioides difficile	ESBL (extended spectrum beta-lactamase) producing *Enterobacterales*		Drug-resistant *Bordetella pertussis*
Carbapenem-resistant *Enterobacterales*	VRE (vancomycin-resistant) *enterococci*		
Drug-resistant *Neisseria gonorrhoeae*	Multidrug-resistant *Pseudomonas aeruginosa*		
	Drug-resistant non-typhoidal *salmonella*		
	Drug resistant *salmonella* serotype Typhi		
	Drug-resistant *Shigella*		
	Methicillin-resistant *Staphylococcus aureus* (MRSA)		
	Drug-resistant *Streptococcus pneumoniae*		
	Drug-resistant *tuberculosis*		

Not only are microbes becoming resistant to single antibiotics, which would be bad enough, but many are becoming resistant to multiple different drugs (MDR or multidrug-resistant). The ease with which bacteria acquire resistance means that in only a few years many bacterial infections may no longer be treatable.

The last completely new class of antibiotics came on the market more than 30 years ago. There are an increasing number of bacteria that are immune or resistant to multiple antibiotics (multidrug-resistant organisms or MDROs). The clinical importance of antibiotic-resistant bacteria varies—but all are important for human health due to the propensity of bacteria to transfer resistance genes both within and between species.

In 2024, the US Centers for Disease Control (CDC) identified five bacterial strains as urgent threats and 11 as serious threats. Another five strains are considered to be concerning or are under watch[1] (Table 2.1).

[1] Antibiotic Resistance Threats in the United States, 2019. 2019 AR Threats Report doi: https://doi.org/10.15620/cdc:82532.

Table 2.2 Antibiotic classes and year of commercial introduction

Year	Antibiotic	Year	Antibiotic	Year	Antibiotic
1935	Sulfonamides	1959	Polymyxins	1971	Phosphonates
1941	β-Lactams	1960	Azoles	1985	Carbapenems
1946	Aminoglycosides	1962	Quinolines	2000	Oxazolidinones
1948	Tetracyclines	1963	Lincosamides	2003	Lipopeptides
1949	Amphenicols	1964	Cycloserines	2007	Pleuromutilins
1952	Macrolides	1964	Cephalosporins	2011	Lipiarmycins
1953	Nitrofurans Tuberactinomycins	1965	Streptogramins	2012	Diarylquinolines
1958	Glycopeptides	1968	Diaminopyrimidines		

Many of these bacteria cause nosocomial (hospital-acquired) infections, although infection in other settings is increasingly common. In the United States alone, more than 2.8 million antibiotic-resistant infections occur each year, and more than 35,000 people die as a result. In addition to these, *Clostridioides difficile* infection (CDI), which is normally not antibiotic resistant but is associated with antibiotic use, affects approximately 500,000 people in the United States and results in about 30,000 deaths (2019). Globally, the numbers are staggering; 4.95 million deaths in 2019 were associated with antibiotic resistance, resulting in 1.27 million deaths.[2]

2.3 Causes of Resistance

The use of antibiotics is the cause of antibiotic resistance. In biological terms, antibiotic treatment is a strong selective pressure. Most bacteria are killed by treatment with antibiotics, but occasionally, a mutation occurs that makes a bacterium resistant. The resistant strain can then spread rapidly since the antibiotic treatment has killed off all competing bacteria. Since the discovery of penicillin in 1928, a steady stream of antibiotic compounds has been discovered or synthesized and brought to market (Table 2.2). However, in recent years, the discovery of new antibiotics has slowed dramatically; only one completely new class of antibiotic has been introduced since 1968, and most "new" antibiotics are modifications or combinations of already known classes.

While any use of antibiotics tends to produce resistance over time, the situation is exacerbated by the widespread overuse of antibiotics. Even though antibiotics are

[2] Antimicrobial Resistance Collaborators, Global burden of bacterial antimicrobial resistance in 2019: a systematic analysis. Lancet 399, pp. 629–655, DOI: https://doi.org/10.1016/S0140-6736(21)02724-0, 2022.

not effective for viral infections, doctors are frequently pressured to prescribe antibiotics for colds and influenza. Before antibiotic resistance was recognized as a widespread problem, it was often felt to be easier, and harmless, to simply give the patient what he/she wanted. The situation is even more serious in many developing countries, where antibiotics can be obtained without a prescription.

Another aspect of overuse is the use of antibiotics to promote growth in food animals when no disease is present. In the late 1990s, about 80% of antibiotics in the USA and Europe were sold for use in livestock, mostly to promote growth in crowded and/or unsanitary conditions. While consumer pressure and new regulations have brought these numbers down, about two thirds of antibiotics worldwide are used in food animal production (and the amount is now increasing). Resistant bacteria have been proven to move from agricultural operations to humans via the food supply and by transmission by workers, to the general community.

Environmental contamination presents yet another route to antibiotic resistance. Confined animal feeding operations produce massive amounts of effluents, often containing remarkably high levels of antibiotics. Hospitals and other urban sources also contribute significant amounts of antibiotic-containing effluent to public sewage systems, exposing large communities of bacteria to selection for resistance.

2.4 Mechanisms of Resistance

Antibiotics have many different mechanisms, but most commonly act by inhibiting enzymes important for cell wall biosynthesis, protein synthesis, transcription, DNA replication, or synthesis of critical metabolites (Table 2.3). Interference with the permeability or integrity of the bacterial cell membrane is another mechanism.

The mechanisms of resistance to antibiotics are as diverse as the mechanisms of the antibiotics themselves, but can be grouped into general categories: decreased uptake/transport of the antibiotic, efflux of the antibiotic or a derivative, breakdown or modification of the antibiotic or a derivative, or mutation/modification of the target molecule. In general, resistance requires only one or a small number of genes. When resistance genes are transferred to episomes (plasmids), small independently replicating circular DNAs found in bacteria, they are very easily transferred between bacteria—even between distantly related species. Drug resistance genes on episomes have been directly linked to many types of antibiotic resistance.

Plasmids, small, independently replicating circles of DNA in bacteria, are directly linked to antibiotic resistance. Typically, once the gene or genes necessary for resistance are transferred to a plasmid, the plasmid spreads rapidly through the population. Plasmids may be transferred between bacteria by conjugation, by transduction (transfer by a bacteriophage), or by release to the environment and direct uptake. Plasmids may be, and frequently are, transferred between species. This transfer of genetic material, by processes other than inheritance from the parent, is

Table 2.3 Antibiotic targets and resistance mechanisms

Mechanism/target	Antibiotic classes	Resistance
Cell wall synthesis inhibitors	B-lactams, cephalosporins, carbapenems, monobactams, vancomycin, bacitracin	B-lactamases, penicillinases Modification of penicillin-binding proteins Modification of cell wall linkage
Protein synthesis inhibitors 30S ribosome subunit 50S ribosome	Aminoglycosides, tetracyclines Macrolides, clindamycin, oxazolidinones, chloramphenicol, streptogramins	Synthesis of ribosomal protection proteins Aminoglycoside-modifying enzymes (AGE) Chloramphenicol-acetyl-transferases rRNA modification Ribosomal protein mutation
Nucleic acid synthesis inhibitors RNA polymerase topoisomerase	Rifampin Fluoroquinolones	RNA polymerase binding protein RNA polymerase mutations RNA polymerase modification Inactivation of my monooxygenase Topoisomerase mutation Transmembrane protein mutation
Folic acid synthesis inhibitors	Sulfonamides, trimethoprim	Dihydropteroate synthase mutation
Decreased membrane permeability/transport Disrupt ion transport Depolarize cell membrane Alteration of cell membrane Porin/membrane import	Ionophores, Lipopeptides Polymyxins, bacitracin	Efflux pumps Phosphatidate cytidylyltransferase mutation Cardiolipin synthase mutation RNA polymerase mutation Phosphoethanolamine transferase mutation LpxACD mutation
Multidrug	All classes	Efflux pumps or inhibition of membrane transporter

called horizontal transfer or lateral inheritance. Many antibiotic resistance plasmids carry resistance to multiple antibiotics (Fig. 2.1), as well as genes that provide the ability to transfer the plasmid to other bacteria. These include genes involved in conjugation or the formation of secretion systems.

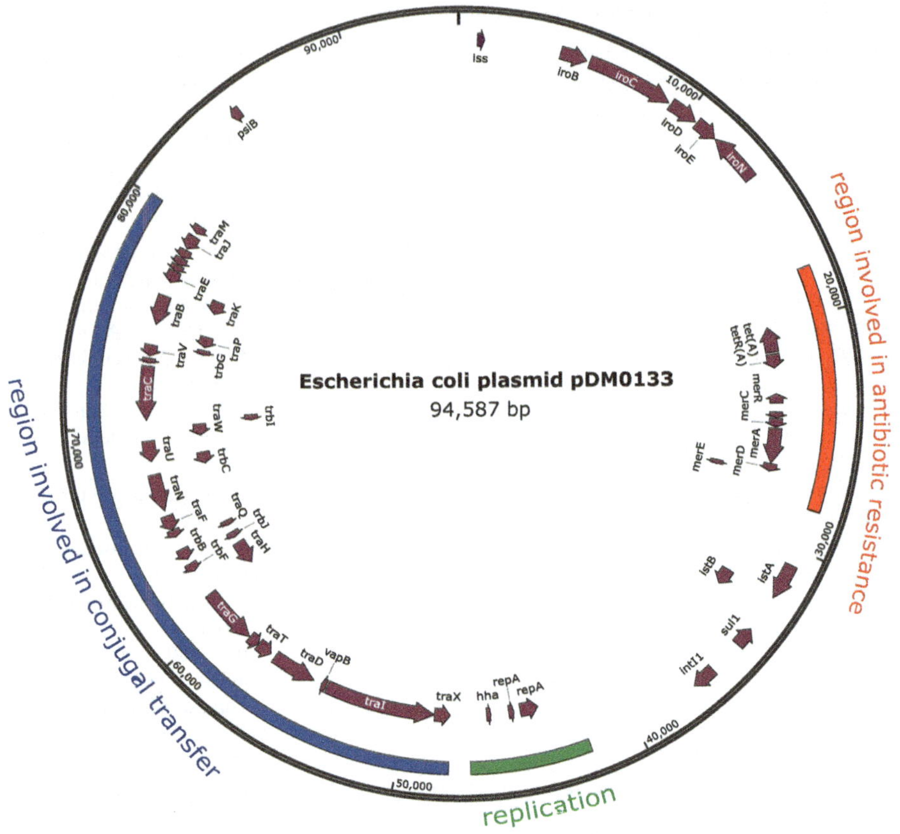

Fig. 2.1 Example of an antibiotic resistance plasmid. Arrows indicate genes. The blue regions encode proteins involved in conjugative transfer. The red genes are involved in antibiotic resistance

2.5 Questions

Using the introductory material and/or online resources, answer the following:

- Why is bacterial antibiotic resistance a serious problem?
- According to the Centers for Disease Control, what are the most dangerous antibiotic-resistant bacteria today?
- What is a nosocomial infection?
- What are the main reasons for that bacteria become resistant to antibiotics?
- What are the main mechanisms of antibiotics?
- What are the main mechanisms of resistance to antibiotics?
- Why/how does antibiotic resistance spread so rapidly?

Further Reading

1. Antibiotic Resistance Threats in the United States, 2019. 2019 AR threats report. 2019. https://doi.org/10.15620/cdc:82532.
2. Antimicrobial Resistance Collaborators. Global burden of bacterial antimicrobial resistance in 2019: a systematic analysis. Lancet. 2022;399:629–55. https://doi.org/10.1016/S0140-6736(21)02724-0.
3. Hutchings MI, Truman AW, Wilkinson B. Antibiotics: past, present and future. Curr Opin Microbiol. 2019;51:72–80.
4. May M. How to fight antibiotic resistance. Nat Med. 2023;29(7):1583–6. https://doi.org/10.1038/d41591-023-00043-5.
5. Overuse of Antibiotics on Farms. http://consumersunion.org/wp-content/uploads/2013/02/Overuse_of_Antibiotics_On_Farms.pdf

Chapter 3
Introduction to Evolution and Homology

Evolution and the relationships between macromolecules, as seen in their nucleotide and protein sequences, are the foundation of sequence analysis. Understanding how the evolutionary process works is key to understanding how we compare sequences and how we identify sequences with an evolutionary relationship. The inheritance of genes and their encoded proteins from common ancestors, with only limited changes due to evolution, means that such genes/proteins (homologs) have similar structure and function. Determining whether gene/protein sequences are homologous is therefore a critical issue in bioinformatics. In turn, determining whether sequences are homologs depends on our confidence that the degree of observed sequence similarity would not be observed by chance. This chapter will discuss the concepts of evolution, homology, and how to determine whether genes/proteins are homologous.

3.1 Objectives

- To understand how evolution causes changes in gene and protein sequences.
- To distinguish genes with a common ancestor from those that are unrelated.
- Understand how to use homology to analyze evolution.

3.2 Two Kinds of Evolution

Evolution requires two simple processes to act together: One process generates inheritable changes in the characteristics of organisms (mutation), and the other removes individuals with low fitness/poor adaptation to the environment (selection). Germline mutations (changes to the genotype of the organism) generate diversity in the characteristics of individuals (phenotypes) in a population by changing the DNA

© The Author(s), under exclusive license to Springer Nature Singapore Pte Ltd. 2025

S. Wang, M. Gribskov, *Practical Bioinformatics*,
https://doi.org/10.1007/978-981-96-7949-2_3

sequence of the genome, however, any process that produces heritable altered characteristics can provide the diversity needed for evolution. Selection, the other essential process, removes individuals in the population based on their individual characteristics. This is usually discussed in terms of an individual's fitness. Individuals with high fitness are more likely to survive and reproduce, thus passing their (possibly altered) characteristics on to the next generation.

Without selection, the characteristics of the individuals in the population gradually become more and more different. This is called *divergent evolution*. Without a mechanism to generate diversity, all individuals in the population are the same and are therefore equally fit. Hence, there is no evolutionary change. In the presence of diversity, selection acts to probabilistically remove less fit individuals, thereby increasing the frequency of characteristics that enhance fitness in the population. This focusing on the characteristics of the population around survival-important traits is *convergent evolution*.

3.3 Homologs, Mutations and the Molecular Clock

Most life on Earth uses DNA as its genetic material. Diversity is generated by mutations in the DNA sequence. Mutations can be generated by many processes, including chemical or radiation damage, incorrect repair of such damage, errors in DNA replication, the action of viruses or transposons that integrate into the DNA, or errors in the process of recombination. Mutation results in four kinds of changes: base substitutions, insertion or deletion of one or more bases, inversions of chromosomal segments, and translocations of DNA segments between chromosomes. Although the mutation process has preferences for certain bases/DNA regions, mutation is relatively random with respect to the information content (the genes and their encoded proteins) of the DNA, and relatively proportional to time. This allows the number of mutational differences between the same gene in different species to be used as a *molecular clock*—the more mutations, the longer the time since the genes diverged from each other. The mutational molecular clock is an example of a stochastic clock. Another example of a stochastic clock is the International Atomic Time (IAT) clock, which is accurate to about one part in 10^{16} (around a billionth of a second per year).

Although stochastic clocks can be very accurate, high accuracy depends on observing a very large number of random events. The IAT clock, for instance, observes a specific electronic transition in Cs-133, which occurs 9,192,631,770/sec. In the case of mutations, there are only a relatively small number of observable events—usually not even thousands. In addition, we cannot observe all the mutational changes—when multiple mutations affect the same base, all we can see when examining sequences from two species is that there has been at least one mutational change. We also do not see mutational changes that are not fixed in the population because the organism with the mutation did not survive and reproduce. These issues make the time estimates derived from molecular clocks very inconsistent (or you can think of them as having a high error rate or variance). Selection, which is not at all random, further complicates the picture by removing particular mutational changes, often in a lineage-specific way.

As evolution proceeds, a single parent species separates into multiple daughter species; the number of mutational differences between corresponding genes in the daughter species can be used to estimate the timing and mechanism of the speciation. Such corresponding genes, which derive from a common ancestor, are called *homologs*. Speciation is one way that genes can be separated and begin to diverge from each other. Another possibility is that a gene can be duplicated, resulting in two *homologous* copies of the gene within a single individual. To distinguish these relationships, we use the terms *orthologs*, genes separated by a speciation event, and *paralogs*, genes separated by a gene duplication event. Homology, similarity in form and function between features in different species resulting from inheritance of a feature from a common ancestor, is a critical concept in bioinformatics. A feature might be, for instance, a protein (in this case, we say the proteins are *homologs*, or that they are *homologous*). Because of their descent from a common ancestor, homologs are likely to have similar structure and function.

3.4 Determine Homologous Sequences

In order to tell if genes or proteins have a common ancestor, and thus can be expected to have similar structure and function, we need to know how similar we expect unrelated sequences to be. A more technical way to express this is to say we need to know the expected distribution of comparison scores for unrelated sequences—when looking at potential homologs, if we observe comparison scores that are much higher than expected for unrelated sequences, we can conclude the genes/proteins we are looking at are related.

When studying an evolutionary process, such as the spread of antibiotic resistance, it is critical that we look at sequences that are homologous—if they are not homologous, the estimates and evolutionary models that we create will be meaningless. Metaphorically, you might think of creating a family tree of your family plus a fish—while there may be some relationship between you and the fish, it is far too distant to make any kind of sensible family tree. Furthermore, in phylogenetic studies, we generally focus on orthologs—paralogs can give misleading estimates of species divergence. For example, consider alpha and beta hemoglobin; these proteins are clearly homologous (have a common ancestor). They are also paralogs (resulting from a gene duplication), since all jawed vertebrates have one or more copies of each. This gene duplication took place about 500 Mya.

Suppose we want to use the mutational change in hemoglobin genes to measure the time of divergence of the human and horse lineages (which we believe to be about 74 Mya), if we use the ortholog pairs, human α-hemoglobin (horse α-hemoglobin) or human β-hemoglobin (horse β-hemoglobin), we should get an estimate close to 74 Mya $\pm$ error. However, if we inadvertently use one of the

```
(a) Global Alignment
Seq A:ATCATCGATAAGCTTAAATGCGGTAGTTTAT
       +        +++++  +++++++++   +
Seq B:-TAGAGTCAGAGCTT--ATGCGGTAGAAAAA

(b) Local alignment
Seq A: AGCTTAAATGCGGTAG
       +++++  +++++++++
Seq B: AGCTT--ATGCGGTAG
```

Fig. 3.1 DNA sequence alignments. The line between the sequences (Seq A and Seq B) shows the positions of identical bases marked with a "+". (**a**) Global alignment that uses the entire length of both sequences. Sequence b contains gaps indicated by "-" that have been inserted to maximize the alignment. (**b**) Local alignment of the same sequences. The local alignment shows the best matching parts of the sequences and omits regions with low similarity

paralogous pairs, either human α-hemoglobin (horse β-hemoglobin), or human β-hemoglobin (horse α-hemoglobin), we will estimate the time as about 500 Mya (the divergence time of α and β hemoglobin), before mammals even existed.

One of the most important questions, therefore, in both systematics and bioinformatics, is determining which genes/proteins are homologous. To start with, we need a way to measure the similarity or distance between genes. For now, we can define our measure of similarity as simply the fraction of identical bases in the aligned sequences of the genes, which is commonly called *percent identity* (Fig. 3.1).

We can also define a distance between sequences as the fraction of nonidentical bases in the aligned sequences (percent difference). It should be clear that S_{ab}, the similarity between genes a and b, and D_{ab}, the distance between genes a and b, are related (Eq. 3.1).

$$S_{ab} = \frac{I_{ab}}{L} \qquad D_{ab} = \frac{L - I_{ab}}{L} = 1 - S_{ab} \tag{3.1}$$

where I_{ab} is the number of identical bases in the aligned sequences and L is the length of the alignment. We will focus on similarity, but keep in mind that similarity and distance are easily converted.

Because different genes and proteins have different evolutionary constraints, they evolve at different rates. In addition, we also do not know the amount of evolutionary time separating genes a and b. This makes it difficult to determine, in advance, what similarity scores to expect for homologous genes. And consequently, it is impossible to give a simple criterion for the degree of similarity needed to conclude that genes are homologous in terms of percent identity. For instance, we CANNOT simply say that genes with greater than 50% identity are homologs.

Sequence comparison scores are based on the sum of the scores for comparison of the pairs of aligned sequence letters (bases/amino acid residues) and negative scores for the introduction of spaces, usually called gaps, needed to maximize the matching. Gaps are marked with "-" in Fig. 3.1. The differences in the aligned pairs of letters correspond to substitutions (missense mutations) in the sequences, and the gaps correspond to insertion-deletion events. Surprisingly, the distribution of comparison scores is similar to the distribution of the length of runs of heads or tails when you flip a coin. This makes it relatively easy to evaluate whether a certain comparison score is likely or unlikely to occur by chance.

It's difficult to predict the exact amount of similarity we expect between homologous sequences because a priori we don't know the divergence time and evolutionary rate for the genes/proteins. On the other hand, we can predict exactly what to expect from unrelated (not homologous) genes. Since such genes have no common ancestry, in the simplest model, we expect the similarity between them to be random, *i.e.*, the same as between any two random sequences. We can calculate the distribution of S_{ab} either analytically (using an equation) or from a simulation (computationally generating random sequences and measuring the similarity). This leads us to make a rule for determining homology that may, at first, seem backward: if sequences a and b are so similar that they are very unlikely to be unrelated, we conclude they are homologous. In statistics, we would state this more formally—the hypothesis that a and b are unrelated is called the null hypothesis, usually denoted by H_0. If we can reject the null hypothesis with a high degree of confidence, we can conclude that a and b are related (homologous). The degree of confidence threshold is commonly denoted α and can be any subjectively determined value; a common choice for α is 0.05 in laboratory experiments, although $\alpha = 0.05$ is usually a very inappropriate value for sequence comparisons (for reasons we will discuss later). Formally, the probability that a measurement is greater than or equal to a critical value (x) is called a P-value. The decision rule is: if the P-value is less than α (Eq. 3.2), we can conclude a and b are homologous (if we choose an appropriate α). P-values must be clearly distinguished from the probability that a measurement will have exactly the specified value $P(S_{ab} = x)$, which is always smaller.

$$P(S_{ab} \geq x) < \alpha \tag{3.2}$$

Sequence similarity is usually measured after the sequences have been aligned, i.e., matched up so that letters in the two sequences represent the ancestral relationship (Fig. 3.1). For two closely related (recently diverged) sequences, you can imagine simply moving sequence A with respect to sequence B at every possible offset and counting the number of identical bases. The offset with the largest percent identity would be the best alignment. The best alignment, therefore, is the result of a maximization of the similarity score across all possible offsets of a and b. For actual biological sequences, the procedure is more complicated since we have to consider adding spaces to sequence A and/or sequence B to account for base/residue insertions and deletions created by the mutational process. These added spaces are conventionally called *gaps*. Because of sequence rearrangements and other large-scale changes in the genome, we may also restrict the alignment to just the most similar (and therefore most likely to be homologous) regions. This kind of alignment, where the less similar regions are trimmed off, is called a *local alignment*. We won't go into the details of how local alignments are computationally determined, but it should be clear that a local alignment is the result of an extensive search over many possible alignments with different offsets between the sequences, with all possible locations, lengths, and numbers of gaps, and different degrees of trimming. The similarity score for the local alignment therefore represents the maximum score out of a huge number of possible alignments.

3.5 Questions

Using the introductory material and/or online resources, define the following:

- What makes a molecular clock an inaccurate timekeeper, in spite of being generated by a random process?
- What is the relationship between the concepts of homology, orthology, and paralogy. Are any of these concepts mutually exclusive?
- What is a germline mutation?
- What are divergent evolution and convergent evolution, and how do they differ?
- What is a *P*-value and how does it differ from a probability?
- What is a global alignment and how does it differ from a local alignment?
- In alignments, what are gaps, and why are they necessary?
- How do the statistics of nonhomologous sequences/species allow us to identify homology?

Further Reading

1. MacIntyre RJ. Molecular evolution: codes, clocks, genes and genomes. BioEssays. 1994;16(9): 699–703. https://doi.org/10.1002/bies.950160918.
2. Saeed U, Usman Z. Biological sequence analysis. In: Husi H, editor. Computational biology. Brisbane: Codon Publications; 2019. Chapter 4.
3. Pearson WR. An introduction to sequence similarity ("homology") searching. Curr Protoc Bioinformatics. 2013;Chapter 3:3.1.1–8. https://doi.org/10.1002/0471250953.bi0301s42.
4. Koonin EV, Galperin MY. Chapter 2—Evolutionary concept in genetics and genomics. In: Sequence—evolution—function: computational approaches in comparative genomics. Kluwer Academic; 2003. https://www.ncbi.nlm.nih.gov/books/NBK20255/#A23.

Chapter 4
Preparing Your Sequences

Scientific information has grown to be very large and detailed. Thousands of organisms have had their whole genome sequence determined, and there are enormous amounts of related data describing gene and protein regulation, predicted and experimentally determined protein structure, protein-protein interactions, gene and protein function, and the relationship of genes/proteins to disease (to mention only a few). Finding information relevant to a specific gene/protein or process can be a long and tedious process. Fortunately, many of the most important repositories are cross-linked to each other—but the sheer volume of information can make finding what you are looking for difficult. In this chapter, we will use the antibiotic colistin as an example and look at a few useful resources to find information about colistin, its mechanism of action, and mechanisms of resistance. Students can follow similar approaches to find information about their antibiotic of interest (aoi).

4.1 Objectives

- Get familiar with some important biological information resources.
- Find information about your antibiotic of interest (aoi).

4.2 Background

Due to the low cost of DNA sequencing, very large resources devoted to scientific literature and to the sequences and structures of both nucleic acids and proteins are publicly available. These resources provide not only the experimentally determined sequences, but also a great deal of additional information on the function, structure, and genetics of biological molecules. This additional information is

© The Author(s), under exclusive license to Springer Nature Singapore Pte Ltd. 2025

S. Wang, M. Gribskov, *Practical Bioinformatics*,
https://doi.org/10.1007/978-981-96-7949-2_4

generically referred to as *annotation*—everything we know about a biomolecule in addition to its sequence.

In this chapter, we introduce some of the most important biological information resources. These resources are publicly available and funded at the national or international level. Most of these resources collaborate with each other to share information, so similar information can be found at multiple resources. As an example, we will look at information related to the antibiotic colistin. Students can follow similar approaches to find information about their antibiotic of interest (aoi). Please keep in mind that biological literature is growing very rapidly, so that the information we report in our examples may have been superseded by newer information by the time this book appears. Keep in mind that the information resources themselves change and adapt as the amount and kind of available information changes, and therefore, their electronic interfaces change as well. The pages displayed at the time you read this text may be somewhat different than our examples here.

The internationally sponsored sequence databases at NCBI, EMBL-EBI, and DDBJ are the *primary repositories* of DNA sequence information, and many scientific journals require authors to deposit sequence information before accepting sequence-related papers. Similarly, protein sequence information is generally deposited in the UniProt database, and three-dimensional protein structure information in the Protein Data Bank (PDB).

- National Center for Biotechnology Information (NCBI). https://www.ncbi.nlm.nih.gov
- European Molecular Biology Laboratory—European Bioinformatics Institute. https://www.ebi.ac.uk/
- UniProt. https://www.Uniprot.org/
- Protein Data Bank. https://www.rcsb.org/

In addition to these primary resources, there are thousands of additional *derived* databases focused on particular topics. A few examples relevant to the activities in this book include:

- Comprehensive Antibiotic Resistance Database https://card.mcmaster.ca
- Genome taxonomy database http://gtdb.ecogenomic.org/

Because of the huge and exponentially growing amount of scientific information, healthcare-related data is growing at about 36% per year,[1] and information management systems such as databases and online applications are essential. Because of the rapid growth, these systems often struggle to keep up with the data flow and to integrate new kinds of data. Nevertheless, basic principles of how to find information tend to remain the same. In this exercise, we will discuss how to design *queries* to find specific information, for instance, about antibiotic resistance.

[1] https://www.rbccm.com/en/gib/healthcare/episode/the_healthcare_data_explosion

The simplest kind of query is a *keyword query*, which searches for a word or phrase in the body of stored information. Keyword queries can be powerful, but if you lack experience in a field, you may find it difficult to identify the correct keywords. For instance, searching the NCBI PubMed literature database for "muscle tumor" produces more than 75,000 results, but inspection of the results shows that many of the *hits* merely mention "muscle" and "tumor" in the same article. Searching for a more technical term, such as "fibrosarcoma," produces almost 20,000 results, which are more closely related to muscle tumors but also include some skin tumors (dermatofibrosarcoma). Finding good keywords usually requires an iterative process of searching, examining the results, and refining your query.

Most database management systems can combine query search terms to produce more powerful queries using the Boolean *AND*, *OR*, and *NOT* operators. Queries with multiple words can be interpreted in different ways; clearly the search for "muscle tumor," above, is interpreted as *muscle OR tumor,* that is, show me all articles containing the words *muscle* or the word *tumor.* Boolean expressions are interpreted left to right (similar to addition), but the order of the operations can be controlled by parentheses (just as in algebra). Here's an example. I know that *tumor* and *cancer* are synonyms, so I want to search for articles and mention *tumor* or *cancer* together with *muscle*. If I search for

muscle AND cancer OR tumor

I get over 4.8 million results, but if I search for

muscle AND (cancer OR tumor)

I only get 131,000. The first query is interpreted as "Show me articles that mention *muscle* and *cancer* in the same article, and all other articles that mention *tumor.*" Whereas the second query is interpreted as "Show me articles that mention *muscle*, and in the same article mention *cancer* or *tumor.*"

One simple extension, used by many databases, allows a phrase to be explicitly searched for by including the query in quotation marks. A search of PubMed for "muscle tumor," with the quotation marks, yields only 1100 results, all of which have the complete phrase (by the way, this approach also works with Google). A few databases also apply *stemming* to convert a specific word, such as *resistance*, and related noun and verb forms, such as *resist, resists, resisted*, or *resisting,* to the common part (the stem), in this case, *resist*. Even if the database you are using does not do this, you can do it yourself manually. Many databases, such as PubMed, allow Boolean operations to be directly included in queries. There is no universal syntax for this; in the exercise below, we will use PubMed as an example.

Many databases, manually or automatically, add additional keywords to their primary data. These additional keywords vary with the database but often include information related to taxonomy, human disease and affected organs, drugs, and more general categories. PubMed extensively annotates entries with MeSH terms (Medical Subject Headings). These are manually determined by expert curators and can be very useful if you are interested in human disease. One benefit of such added information is that it follows a standardized vocabulary, making it easier to create queries (the MeSH vocabulary includes about 30,000 terms, so learning it is not trivial).

4.3 Before Lab Questions

- Why does a PubMed search for "colistin resistance" produce different results than "colistin AND resistance"?
- How would you search for articles that mention colistin, and antibiotic resistance but not polymyxin (polymyxins are the class of antibiotics that include colistin)?

4.4 Resources and Materials

- PubMed—https://pubmed.ncbi.nlm.nih.gov/
- UniProt—https://www.Uniprot.org/
- Comprehensive Antibiotic Resistance Database—https://card.mcmaster.ca/
- NCBI databases—https://www.ncbi.nlm.nih.gov/

4.5 Activity 1: Selecting an Antibiotic of Interest (aoi)

In order to provide practical examples of different bioinformatics methods, we use the polymyxin antibiotic colistin as our example. Using the background information on antibiotics (Chap. 2), and particularly in Tables 2.1, 2.2, and 2.3, students should select an antibiotic to focus on in the exercises in this book. We will refer to this as your antibiotic of interest (aoi). Since we use colistin as an example, you should not choose colistin or polymyxins in general as your aoi. In addition to the background information included in this book, students can use internet searches to find current news, journal articles, and other information that will help them choose an interesting aoi. Wikipedia (https://www.wikipedia.org/) is a particularly useful source of basic information.

Notebook
As the first entry in your notebook, record your aoi and the preliminary information that you have found. It is important not just to cut-and-paste information in your notebook—be sure to record the source of the information as well. If information contains a date, this is relevant since scientific information is frequently updated.

4.6 Activity 2: Finding Some General Information About Your aoi

A good place to begin is by finding a fairly recent review article discussing your antibiotics. Review articles tend to be easier to read because they are intended for a more general audience, and they summarize information from a range of more detailed papers. In the instructions below, we use the abbreviation aoi to indicate your antibiotic of interest. We use colistin as an example.

Step 1: Go to NCBI PubMed. Compare the number of results you see for the following queries, replacing aoi with the name of your chosen aoi:

- aoi resistance
- aoi AND resistance
- "aoi resistance"

> **Notebook**
> Record the number of results you see for each query. Examine the results and discuss in your notebook which search results you think are most relevant and why. How is the query in quotes interpreted differently?

Step 2: Our results for colistin are shown below (Fig. 4.1). There are thousands of results to look at, so we will restrict the search in two ways. First, we will limit the

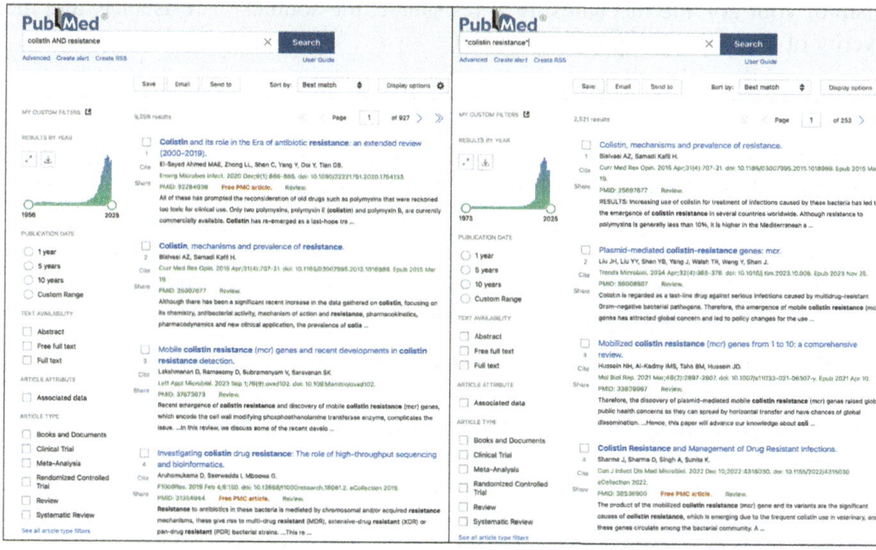

Fig. 4.1 Examples of PubMed searches (https://pubmed.ncbi.nlm.nih.gov). The left panel shows a search using Boolean operator. The right panel shows matching to a phrase using quotes

results to just review articles, which can be selected using the ARTICLE TYPE checkboxes at the left. Second, we will restrict the range of dates to only those papers from 2020 to 2024 using the slider on the graph on the upper left. This still yields over a thousand results. Scanning through the article titles, I can see that many of the articles are about specific outbreaks. To find articles that are more focused on the mechanism of resistance, I try a more specific search:

"colistin resistance" AND mechanism

Now I get only 569 results. Scanning down the results, I immediately see three that sound like they would be informative:

1. Molecular mechanisms of colistin- and multidrug-resistance in bacteria among patients with hospital-acquired infections. Khurana S, Katiyar A, Puraswani M, Sharma D, Walia K, Malhotra R, Mathur P. Future Sci OA. 2023 Aug 28;9(10):FSO896. https://doi.org/10.2144/fsoa-2022-0055.
2. Plasmid-mediated colistin-resistance genes: mcr. Liu JH, Liu YY, Shen YB, Yang J, Walsh TR, Wang Y, Shen J. Trends Microbiol. 2023 Nov 25: S0966-842X(23)00295-0. https://doi.org/10.1016/j.tim.2023.10.006.
3. A Review of Resistance to Polymyxins and Evolving Mobile Colistin Resistance Gene (mcr) among Pathogens of Clinical Significance. Shahzad S, Willcox MDP, Rayamajhee B. Antibiotics (Basel). 2023 Nov 6;12(11):1597. https://doi.org/10.3390/antibiotics12111597.

Step 3: Perform similar searches for your aoi. Find articles that describe the mechanism of your aoi, the mechanisms of resistance, the sources of resistance, and the severity of the resistance problem.

Notebook

Include the 2–3 most informative references in your notebook, as in the example above. Summarize the information in several paragraphs and include the text in your notebook. Pay particular attention to identifying the gene(s) involved in resistance. Keep in mind that many drugs/genes/proteins have synonyms that may be important in identifying relevant information (record these as well). You should be able to find information about:

- What are gene names and functional synonyms for the proteins in this family?
- What antibiotics does the protein family provide resistance to?
- How do the proteins cause resistance?
- Is there an experimentally determined of prediction structure of a resistance protein in this family available ?
- How widespread is the resistance?
- Is resistance a serious medical problem?
- How many patients are affected per year?

Below is an example of summarized information about colistin.

Colistin is the antibiotic of last resort for multiple drug resistant Gram-positive bacterial infections. Colistin belongs to a class of drugs called polymyxins, which were discovered in 1947 and went out of use in the 1950s due to problems with nephrotoxicity. However, as the number of antibiotics available to treat multiple antibiotic-resistant bacteria declined in recent decades, polymyxins were returned to use. Unfortunately, in November 2015, colistin resistance genes were identified in a large number of samples in China, probably associated with the use of Colistin on pig farms, and have since been identified around the world.

Colistin is a cationic cyclic polypeptide that acts by interacting with lipopolysaccharides (LPS) and phospholipids in the outer membrane of Gram-positive bacteria. It displaces divalent cations, resulting in disruption of the cell membrane, and, eventually, cell lysis and death. The most common mechanism of colistin resistance involves altering the LPS in the outer cell membrane so that it no longer interacts with colistin. Two of the known colistin resistance genes, mcr-1 and mcr-2, are classified as phosphoethanolamine transferases and are believed to act by modifying the membrane LPSs (Fig. 4.2).

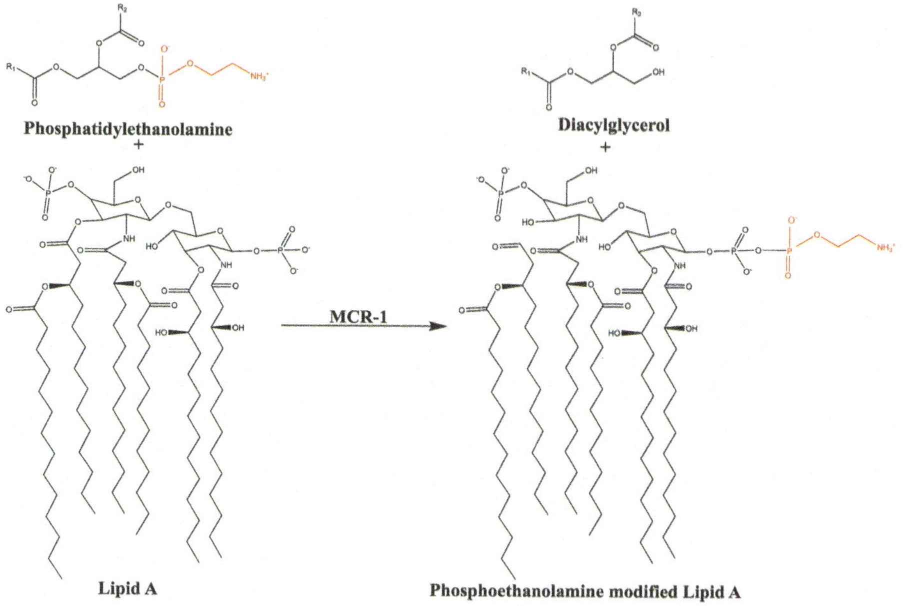

Fig. 4.2 Chemical structure of mcr-1

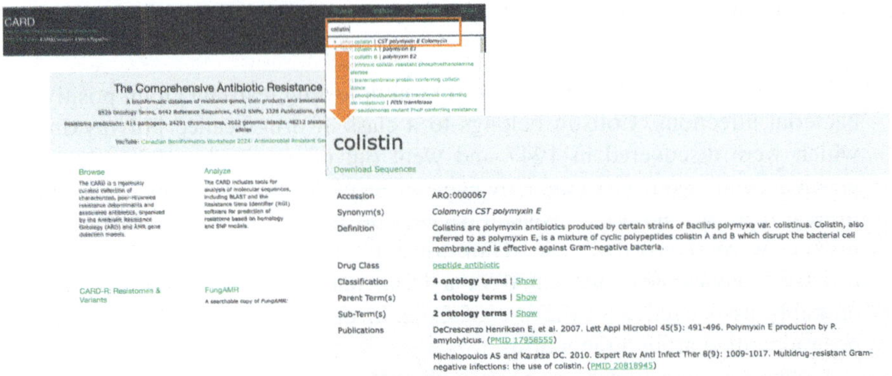

Fig. 4.3 CARD database search (https://card.mcmaster.ca) for colistin

4.7 Activity 3: Obtaining Antibiotic Information from CARD Database

The Comprehensive Antibiotic Resistance Database[2] (CARD) has a large collection of genes and mutations associated with specific antibiotics. CARD can be used to predict antibiotic resistance based on genomic sequence, predict resistance in previously unknown mutants, and to design probes for detecting resistance genes. We will use CARD to find more information on your aoi.

Step 1: Enter the name of your antibiotic (Fig. 4.3), we'll us colistin as an example, into the *search* box on the upper right. Each item in the list provides information about a kind of resistance, including it origin, gene family, drug class, resistance mechanism, and literature references.

Step 2: Scanning through the entries in this list, you should be able to come up with a pretty complete list of the modes of resistance for your aoi.

Notebook

Search for your aoi. If you found multiple genes and modes of resistance, this will require multiple searches. Summarize related information and carefully choose genes that are specific to your aoi. For example, the emrA gene, which provides some resistance to colistin is not specifically related; EmrA is part of an antibiotic efflux pump, which gives resistance to many antibiotics but primarily to fluoroquinolones not polymyxins.

---

[2]Alcock et al. 2023. CARD 2023: Expanded Curation, Support for Machine Learning, and Resistome Prediction at the Comprehensive Antibiotic Resistance Database. Nucleic Acids Research, 51, D690-D699

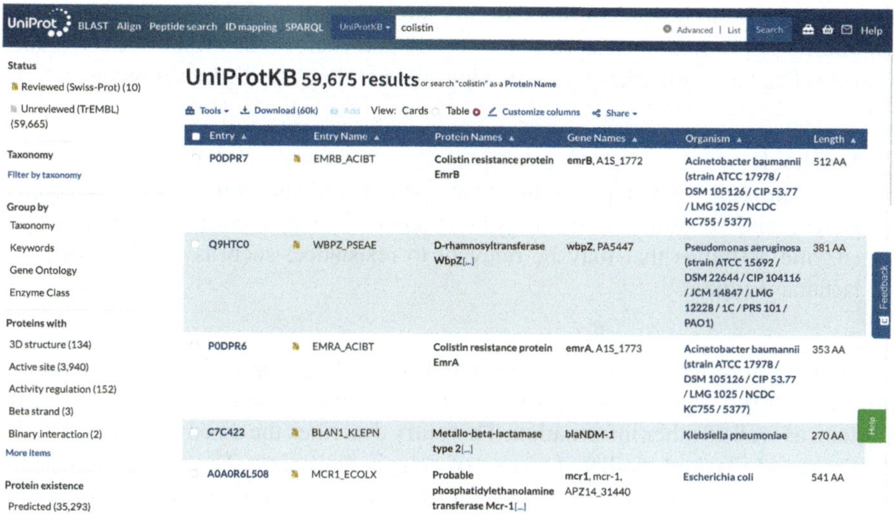

Fig. 4.4 UniProt search (https://www.Uniprot.org/) for colistin

4.8 Activity 4: Obtaining Sequence Information from UniProt

UniProt,[3] is a major international database for proteins, supported by both US and European funding agencies. It is a collaboration between the European Bioinformatics Institute (EMBL-EBI), the Swiss Institute of Bioinformatics (SIB), and the Protein Information Resource (PIR). In this activity, we will leverage some of the diverse information at UniProt to find out more about your aoi.

In the previous activities, we already found some information related to colistin. These include the general class of the antibiotic (polymyxin), resistance genes (mrc-1 and mcr-2), and the formal name of the enzyme involved in resistance (phosphoethanolamine transferase). These make excellent queries for the UniProt database.

Step 1: Searching UniProt produces lists of entries describing genes (Fig. 4.4). Examples of gene entry names are Q9HTC0 or WPBZ_PSEAE.

Simply by looking at the first page, I can see that I have found additional genes that may be involved in colistin resistance (for instance, emrA and emrB, and blaNDM-1). Since genes are frequently given different names in different organisms, it isn't clear that these represent functionally different genes; more information is required to figure out how many distinct genes may be involved. One important clue is the length of the protein. Homologous proteins with similar functions tend to

[3]The UniProt Consortium, UniProt: the Universal Protein Knowledgebase in 2025, *Nucleic Acids Research*, Volume 53, Issue D1, 6 January 2025, Pages D609–D617.

be about the same length. In this case, the protein lengths fall into several groups: around 510–540 amino acid residues (emrB and mcr-1), or around 350 residues (emrA). This large difference in size suggests that they are distinct proteins.

Notebook
Look through several pages of results and record the gene names, protein names, and protein lengths in your notebook. Also, record any additional enzyme activities that may be relevant to resistance, such as "metallo-beta-lactamase type 2".

Step 2: More information can be found by clicking the Entry ID (for instance A0A0R6L508). The text will provide some information about the function of the protein, as well as other information. This entry describes the discovery of mrc-1, the first reported plasmid-mediated source of colistin resistance. The text also provides literature references to more in-depth information. Much more information is available, including recommended gene and protein names, matches to family and domain databases, and the amino acid sequence itself (downloadable).

We will use all this information in later exercises. The entry also provides an array of links for downloading the gene sequence and the protein sequence from EMBL/GenBank/DDBJ. The list of similar proteins is particularly useful in extending the list of genes/proteins involved in this kind of resistance. Also note that there are links to both predicted and experimentally determined three-dimensional protein structures (for colistin, there are more than 10 structures).

Notebook
Familiarize yourself with the kinds of information available on UniProt and update the synopsis describing your aoi.

4.9 Activity 5: Download Sequences from NCBI

From previous activities, you already have some knowledge of your aoi, we can obtain further information about the resistance genes of your aoi. The first thing we need is the DNA sequence and protein sequence of the genes. Later in the following chapters, we will use these sequences to analyze and identify other relevant information, such as homologous genes, protein domains, and three-dimensional structure.

Step 1: Go to the NCBI nucleotide database, and enter the gene name of your aoi in the search box; in this case, we used mcr-1 as an example (Fig. 4.5).

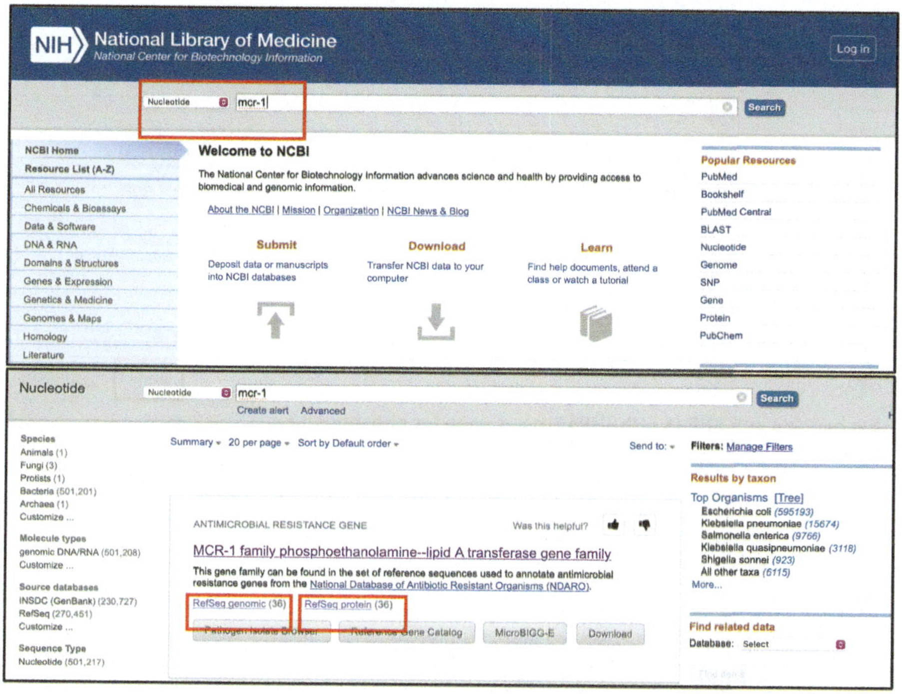

Fig. 4.5 Search mcr-1 DNA and protein sequences from NCBI database (https://www.ncbi.nlm.nih.gov/)

Step 2: The returned results will give you information about the mcr-1 gene (Fig. 4.5).

Step 3: Clicking the RefSeq genomic link will direct you to the DNA sequence (Fig. 4.6). You can see that there are many organisms that have the mcr-1 gene. Since we are interested in antibiotic resistance, I chose the first entry, which describes a sequence from E. coli, a common gut bacterium which can be pathogenic, and click the FASTA link to download the sequence in FASTA format.[4]

Step 4: Now we can see the whole DNA sequence of the mcr-1 gene. Under "Choose Destination," select "File" and choose FASTA format. Finally, select "Create File" (Fig. 4.6). The DNA sequence will be downloaded to your computer.

Step 5: Repeat steps 3 and 4, but, this time, click "RefSeq protein" (Fig. 4.5). You will be directed to the protein sequence page (Fig. 4.7). Again, we choose an entry for *E. coli* and click FASTA, then download protein sequence.

---

[4]FASTA format is a simple format that is the *lingua franca* of bioinformatics. It consists of a title line beginning with ">," followed by the sequence ID (ending at the first space). Anything else on the title line is documentation describing the sequence. After the title line, all remaining lines are the nucleotide or amino acid sequence.

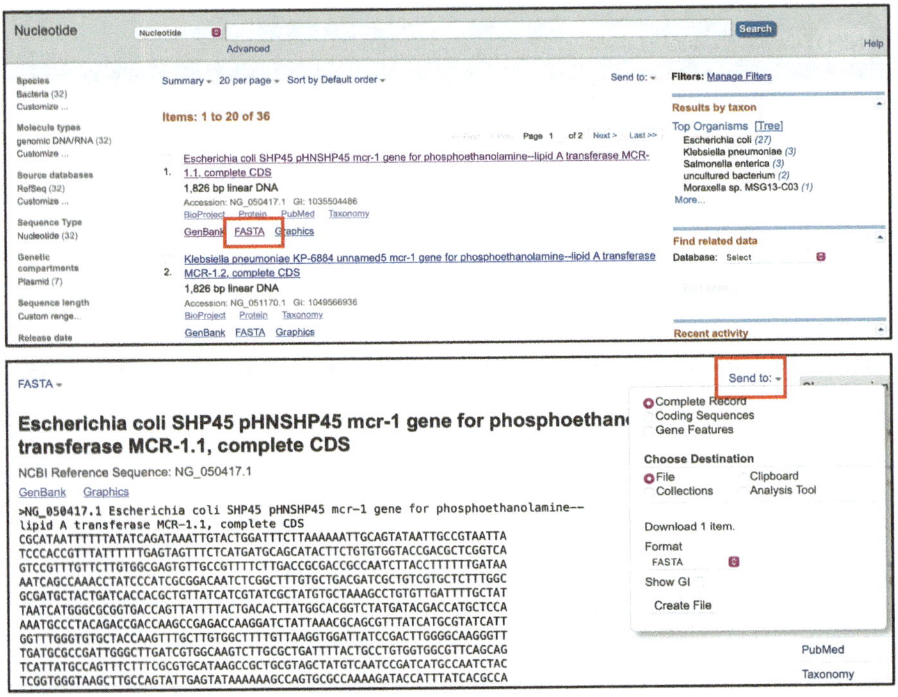

Fig. 4.6 Download mcr-1 DNA sequence from the NCBI database (https://www.ncbi.nlm.nih.gov/)

Notebook
Search for sequences of your aoi (DNA and protein), and record which organism you choose and the sequence information, for example, sequence IDs, and the lengths of the DNA and protein sequences. Make sure you save them on your computer, and we will use them in the later chapters.

4.10 Analysis/Notebook

From activity 1 to activity 5, you will see a lot of information that is cross-linked and repeated in the results. Please summarize the information of your aoi in several paragraphs and organize the information in a clear structure so you can easily retrieve the information later.

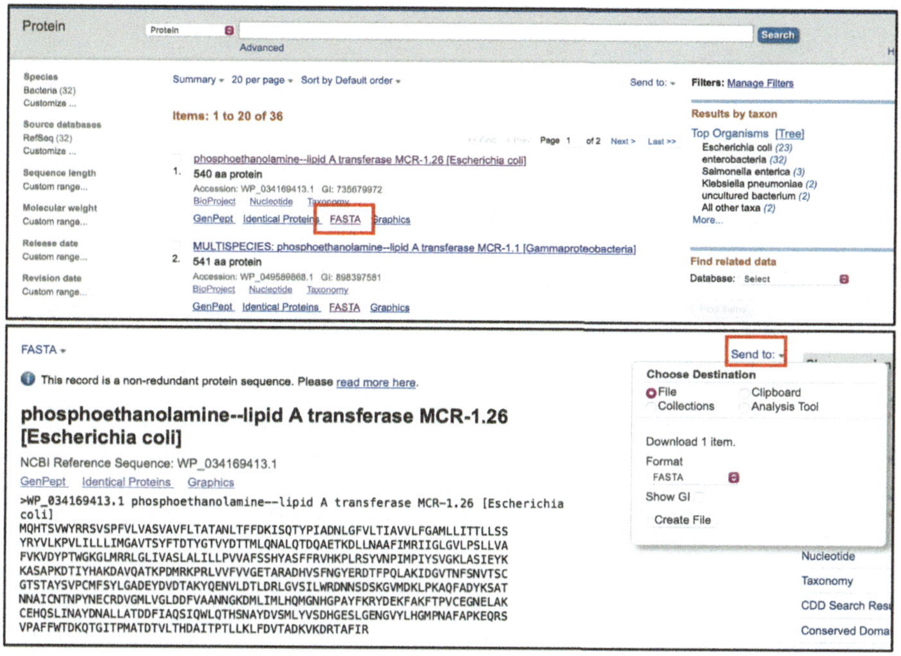

Fig. 4.7 Download mcr-1 protein sequence from NCBI database (https://www.ncbi.nlm.nih.gov/)

Further Reading

1. For details on building PubMed queries: https://pubmed.ncbi.nlm.nih.gov/help/

Chapter 5
Investigating Sequence Comparison Statistics

From previous chapters, students have already obtained the sequences of the genes and proteins related to their aoi. In order to obtain more information about these genes/proteins, such as their evolutionary relationships, it is useful to obtain the gene/protein sequences of related genes/proteins. This is important because the conserved regions of homologs are closely related to the structure and function of the molecules. To determine if genes are homologous, we need to determine if their sequences are more similar than we would expect them to be by chance. When we compare two sequences, we measure how similar/different the sequences are. Our conclusion about whether the genes/proteins are homologous are drawn based on hypothesis testing. We propose the null hypothesis that these sequences are random vs the alternative hypothesis that these sequences are related. Therefore, it is important to understand the statistical background of sequence comparison. Normal distributions are often used in statistical analyses, however, for sequence comparison, it follows an extreme value distribution (EVD), which we will learn in this lab.

5.1 Objectives

- Understand the concept of extreme value distributions.
- Know the properties of EVD and why we use EVD in sequence comparison.
- Understand the differences between EVD and normal distributions.

5.2 Background

In order to decide if sequences are homologous, we need to know what similarity scores to expect for unrelated sequences. Obviously, some random sequences will be more similar, while others will be less similar, creating a distribution of expected

© The Author(s), under exclusive license to Springer Nature Singapore Pte Ltd. 2025

S. Wang, M. Gribskov, *Practical Bioinformatics*,
https://doi.org/10.1007/978-981-96-7949-2_5

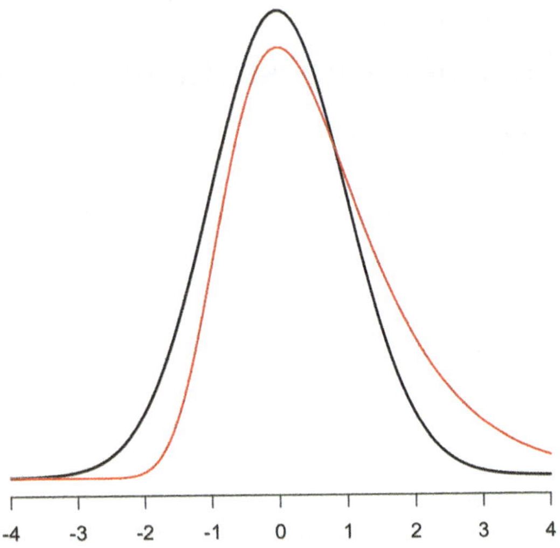

Fig. 5.1 Distributions. Normal distribution is shown in black and an extreme value distribution in red. The X axis units are the number of standard deviations from the mean of the normal distribution

scores. While many biological measurements follow a *normal* distribution, making evaluation of *P*-values simply a lookup in a table or a press of a calculator button, the distribution of values from a maximization follows a different distribution. The distribution of maxima follows an *extreme value distribution* (EVD), also known as a *Gumbel distribution*. An EVD has a larger tail on the positive side than a normal distribution. This means that if you use estimates of significance (*P*-values) based on a normal distribution, you will greatly overestimate the significance of gene similarity based on an alignment score.

One example might be plotting a histogram of the height of all of the students at Purdue. If we gather enough measurements and plot them as a histogram, we would expect a normal distribution like the black curve in Fig. 5.1. However, if we go to each classroom at Purdue and record only the tallest person's height, that is the maximum height in a certain population, we would expect the distribution to be more like the red curve in Fig. 5.1. Maximizations always result in EVDs.

5.3 Materials

- Two index cards
- A fair coin
- Data file: chapter5_sequence_data.xls (supplemental file 1)

5.4 Activity 1: Coin Flipping

The purpose of this activity is to see why extreme value distributions occur and compare what you see with random and nonrandom data.

When a person flips a coin (the coin has two sides, one named heads and one named tails), we expect that he/she will throw a tail 50% of the time and a head 50% of the time. In this activity, we will produce two data sets. We call one dataset the *fake* dataset; this dataset is generated by the student by simply writing down a series of heads and tails. We call the other dataset the *real* dataset, which is produced by actually flipping a coin. We will compare the differences between these two data sets. The challenge is: can you distinguish between the fake and real datasets by looking at the series of heads and tails?

Step 1: Generate the fake dataset.

It is easiest if students work in pairs. With one person acting as a scribe, have the other person make up a random series of 100 heads and tails out of their head. The scribe will write these down as a series of letters "H" and "T" on an index card. In order to make the following analysis easier, please write your data in groups of five letters separated by a space as shown below:

HHTTH TTHHT HHTTH HHTTT ....

On the reverse side of the card, write the name of your group, and "fake" so that you know this is the fake data.

Notebook
Record the fake dataset in your notebook.

Step 2: Produce real data.

Next, we'll generate a set of "real" data by having one person actually flip a coin 100 times. Write down the series of "H" and "T" on another index card with the same format (five letters in a group). Mark the reverse of the card with the name of your group and "real."

Notebook
Record the real dataset in your notebook.

Step 3: Determine the number of heads per length = 5 window. For both the real and fake data, count the number of heads in each window of 5 and include the result on your index card as tables (Fig. 5.2). Now you see why it is easier if you write the H/T in groups of five. There is no maximization here, so we expect the frequencies to be binomially distributed (the binomial distribution is a bell-shaped curve that converges to a normal distribution if the window is large and there are sufficient observations). Because we only generated 100 long sequences, we see a lot of random variation.

	Fake Data	Number	Real Data	Number
1	TTHTH	3	HTTTH	3
2	HTHHH	1	TTHTH	3
3	TTHTH	3	HHTHT	2
4	HHTTH	2	THHTH	2
5	TTTHT	4	THHTT	3
6	HHTTT	3	HHTTH	2
7	HTHHT	2	TTHHT	3
8	THHTH	2	HHHHT	1
9	TTHHT	3	TTTTT	5
10	HHHTT	2	TTHTH	3
11	HTHTT	3	HTTTT	4
12	HHHHT	1	HTTTH	3
13	THTHT	3	THTHT	3
14	THTHT	3	TTHHT	3
15	THHTH	2	TTTTH	4
16	TTHTT	4	HHTHT	2
17	THTTT	4	HHTTT	3
18	HTTHH	2	TTTTH	4
19	THHHT	2	HHHTT	2
20	HHHTH	1	HHTTT	3

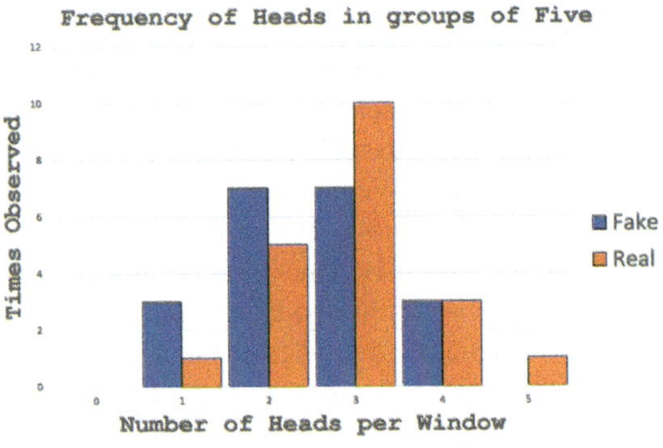

Fig. 5.2 Frequencies of heads in groups of five

Optional At this point, students may exchange cards and see if they can distinguish the *real* and *fake* data without looking at the label on the reverse. Alternatively, turn the cards into the instructor and challenge the instructor to sort the cards into real and fake groups.

Turn in your index card and the instructor will add up the data from all groups to produce a larger dataset. Plot the histograms of the initial data and the summed data, for both real and fake datasets.

Notebook
Save the frequency tables and histograms for your real and fake datasets. Plot the histograms for the summed data (provided by the instructor). Can you distinguish the real and fake datasets, in either the initial or summed data, based on this analysis. Why or why not?

Step 4: Determine the lengths of runs of heads and tails.

A run of heads or tails is the number of consecutive heads or tails thrown. For instance, for the series of coin tosses HTTHH HHTHH, there are two runs of length 1 (1H and 1 T), two of length 2 (1 HH and 1 TT), and one of length 4 (HHHH). Report the number of runs of each length as a table with columns showing the length of the run and the number of times it was observed (Fig. 5.3). Find the length of the

(a)

Fake Data			
TTHTH	HTHHH	TTHTH	HHTTH
TTTHT	HHTTT	HTHHT	THHTH
TTHHT	HHHTT	HTHTT	HHHHT
THTHT	THTHT	THHTH	TTHTT
THTTT	HTTHH	THHHT	HHHTH

Real Data			
HTTTH	TTHTH	HHTHT	THHTH
THHTT	HHTTH	TTHHT	HHHHT
TTTTT	TTHTH	HTTTT	HTTTH
THTHT	TTHHT	TTTTH	HHTHT
HHTTT	TTTTH	HHHTT	HHTTT

(b)

Frequency table		
Length	fake	real
1	31	22
2	19	14
3	9	6
4	1	3
5	0	1
6	0	0
7	0	1
8	0	1

(c)

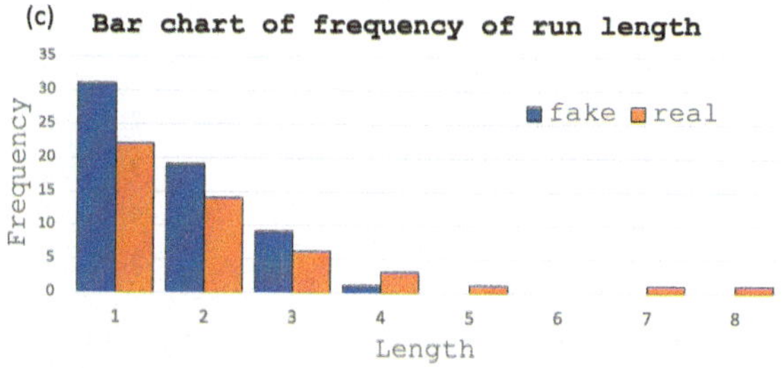

Fig. 5.3 Frequency table and graph of different length of runs. (**a**) Examples of fake and real data. The run length is indicated by color. Length 1 is red, length 2 is blue, length 3 is orange, length 4 is green, length 5 is purple, length 7 is pink, length 8 is black. (**b**) The frequency table of different lengths in fake and real data. (**c**) Bar graph of the frequency of run-lengths

longest continuous run of heads or tails for both fake and real data, and write the number at the bottom of the front side of the respective index cards (the side where you recorded the series of heads and tails). This is a maximization process, so we expect that the length of the longest run should reflect an EVD.

Notebook
Include the table of run lengths, and plot histograms of the distributions of the run lengths. Discuss the differences that you see between the real and fake datasets.

5.5 Activity 2: Real DNA Sequences Showing the Effect of Sample Size

There are two main reasons why it is hard to distinguish between the real and fake data in Activity 1. First, we looked at a very small number of observations, only 100. Second, the window we used is small, making the approximation to a normal distribution difficult to see. In this activity, we will look at the same kind of data, but using much larger computationally generated samples (supplemental file 1).

We are interested in two statistics: the number of matches between sequences and the maximum number of consecutive matches (called run-length, above).

(a)

```
seq1A: CTAGTTTCAGTTGGG
       | |  |      ||
seq1B: CAACCTGTGAGCGGA

seq2A: ATGCTAGTGCAATGG
        || ||
seq2B: TTGGTACGATTCAAA

seq3A: TCCTACGGTTGGAAT
       |||||
seq3B: TCCTATCAACTCTGA
```

(b)

	# of Matches	# of maximum length
Pair 1	5	2
Pair 2	4	2
Pair 3	5	5

(c)

Frequency table of numbers		
Number	Match	Maximum length
0	0	0
1	0	0
2	0	2
3	0	0
4	1	0
5	2	1

Fig. 5.4 Converting an alignment into a frequency table. (**a**) Three representative pairwise alignments. (**b**) Match and maximum run-length scores for the representative alignments. (**c**) Frequency table of the count (number) of matchesand maximum-run scores summed across the representative alignments A, B, and C

Remember, we are interested in random sequences because if we can confidently say that two sequences are NOT RANDOM with respect to each other, then they must be related (homologous).

Consider a simple example. Suppose we have three pairs of sequences to compare (Fig. 5.4a). We can count the number of matches for each pair and get a count table (Fig. 5.4b). Then we can get the frequency for all the numbers (Fig. 5.4c), in this simple case, they are 4,5,5 for number of matches and 2,2,5 for maximum length.

Students may expect that two unrelated sequences would have a percentage identity of zero. However, any pair of sequences will be randomly matched at some positions. The data we are using comes from computationally generated 1000-base-long random sequences. We performed pairwise comparisons, and for each pairwise comparison, we determined the number of matched bases and the maximum consecutive match between the sequences. Examples of the frequency tables of the counts are shown in Fig. 5.5. This frequency table is stored in supplemental file 1.

We will the data in the first sheet (called size) to investigate how the sample size influences the distribution. We simulated four sample size from 1000 pairs to 1,000,000 pairs (Fig. 5.5).

Students may notice that there are always some matches. Even though these sequences are completely random the number of sequence-pairs with zero matches is always zero. For sequences that have equal numbers of the four bases, such as those shown here, the number of random matches should be about 25% of the sequence length; this is why no matches are seen in the truncated table shown in Fig. 5.5.

	A	B	C	D	E	F	G	H	I	J	K	L
1	50% GC 1000 pair of comparisons			50% GC 10000 pair of comparisons			50% GC 100000 pair of comparisons			50% GC 1000000 pair of comparisons		
2	number	match	max length	number	match	max length	number	match	max length	number	match	max length
3	0	0	0	0	0	0	0	0	0	0	0	0
4	1	0	0	1	0	0	1	0	0	1	0	0
5	2	0	0	2	0	0	2	0	0	2	0	0
6	3	0	0	3	0	0	3	0	0	3	0	4
7	4	0	54	4	0	519	4	0	5187	4	0	52060
8	5	0	417	5	0	4318	5	0	42798	5	0	428814
9	6	0	327	6	0	3475	6	0	35439	6	0	352565
10	7	0	145	7	0	1238	7	0	12172	7	0	121814
11	8	0	42	8	0	335	8	0	3239	8	0	33095
12	9	0	11	9	0	88	9	0	874	9	0	8690
13	10	0	3	10	0	19	10	0	218	10	0	2241
14	11	0	0	11	0	6	11	0	54	11	0	533
15	12	0	1	12	0	2	12	0	15	12	0	131
16	13	0	0	13	0	0	13	0	2	13	0	39
17	14	0	0	14	0	0	14	0	2	14	0	10

size GC +

Fig. 5.5 The frequency table of the number of matches and maximum run-length for different sample sizes with 1000 bases long DNA sequences

> **Notebook**
> Plot histograms of the number of matches for the different size samples, and also plot histograms of the distribution of maximum match length for the different sample sizes. See Figs. 5.2 and 5.3, for examples. What is the effect of increasing the sample size (pairs of sequences) on the histograms.

5.6 Activity 3: Real DNA Sequences Showing the Effect of Sequence Composition

One of the major variables that affects the number of matches and run-lengths is the overall GC content of the sequences. The more skewed the AT vs GC composition is, the more a sequence resembles a sequence of just two letters instead of four. Intuitively, if the difference in GC and AT content is greater, we should see more matches between random sequences, and we should see longer maximum consecutive matches.

Using the provided data file called "GC" in supplemental file 1 (Fig. 5.6), plot histograms of the numbers of matches per window and the maximum run length. The data were randomly generated for a large number of sequences, each 1000 bases long. The random generation was repeated for different GC contents so that you can see the effect of sequence composition.

> **Notebook**
> Use the data in the second Excel worksheet titled "GC." Plot histograms of the number of matches for the different GC content, and also plot histograms of the distribution of maximum run length for the different GC content. Does the

(continued)

10% GC			20% GC			30% GC			40% GC			50% GC		
number	match	max length	number	match	max length	number	match	max length	number	match	max length	number	match	max length
0	0	0	0	0	0	0	0	0	0	0	0	0	0	0
1	0	0	1	0	0	1	0	0	1	0	0	1	0	0
2	0	0	2	0	0	2	0	0	2	0	0	2	0	0
3	0	0	3	0	0	3	0	0	3	0	0	3	0	0
4	0	0	4	0	10	4	0	629	4	0	3378	4	0	5187
5	0	92	5	0	4889	5	0	22636	5	0	38513	5	0	42798
6	0	5723	6	0	31376	6	0	42600	6	0	37948	6	0	35439
7	0	25913	7	0	34590	7	0	22726	7	0	14507	7	0	12172
8	0	30827	8	0	18150	8	0	7976	8	0	4147	8	0	3239
9	0	19971	9	0	7058	9	0	2443	9	0	1109	9	0	874
10	0	9987	10	0	2628	10	0	688	10	0	286	10	0	218
11	0	4315	11	0	881	11	0	229	11	0	82	11	0	54
12	0	1872	12	0	270	12	0	59	12	0	19	12	0	15
13	0	754	13	0	94	13	0	12	13	0	10	13	0	2
14	0	344	14	0	37	14	0	2	14	0	0	14	0	2

Fig. 5.6 Screenshot of the frequency table of number of matches and maximum length for 1000 base DNA sequences with different %GC content (comparison of 100,000 sequence pairs)

change in GC content change the peak in either the match histogram or the run-length histogram?

5.7 Analysis/Notebook

- In activity 1, you plotted histograms of the number of heads in a window of five, in what ways do the two distributions (fake and real data) look different? How about the run-length distributions? Are the histograms for fake and real data also different? Record your observations, histograms, and the answers to questions in your Notebook.
- In activity 2, what is the effect of sample size on the histograms?
- In activity 3, describe the effect of varying GC content on the distribution of matches and run lengths. Does the skewed composition affect the symmetry of the distributions? How much more likely is a run of 10 or more in 10% GC sequences compared to 50% GC sequences?

Further Reading

1. Koonin EV, Galperin MY. Chapter 2: Computational approaches in comparative genomics. In: Sequence – evolution – function. Boston: Kluwer Academic; 2003. https://www.ncbi.nlm.nih.gov/books/NBK20260/.
2. Mitrophanov AY, Borodovsky M. Statistical significance in biological sequence analysis. Brief Bioinform. 2006;7(1):2–24. https://doi.org/10.1093/bib/bbk001.

Chapter 6
Finding Homologous Sequences Using BLAST

One of the most common bioinformatics analyses is to identify proteins that are homologous to a sequence of interest (soi). Sometimes the soi is a novel protein from a new genome, and we want to identify homologs in order to predict its function. In other cases, we may want to know how widespread proteins with a specific function are. In this chapter, we search for homologs of proteins that cause resistance to *colistin*, which was until recently one of the few antibiotics without widespread resistance (no longer true, unfortunately). Students should have chosen an aoi in Chaps. 2 and 3, and use the same approach to identify homologs of your aoi. We will use the basic local alignment search tool (BLAST) to identify homologs of your aoi.

6.1 Objectives

- Understand how the similarity of amino acid residues is defined and used to score sequence comparisons.
- Understand the main steps in the BLAST database searching programs.
- Know when to apply the various BLAST programs: BlastN, BlastP, BlastX.
- Apply BLAST searches to identify genes related to antibiotic resistance.

6.2 Background

With the rapid advances in sequencing technology, many individual genes have been sequenced, and many organisms have had their complete genomic sequence determined. Because this information is extensive, we store sequence information in databases. When we obtain a new sequence, we often search the databases to identify homologous sequences or genes. If we are confident that the previously known sequences in the database (subject sequences) are homologous to our query

© The Author(s), under exclusive license to Springer Nature Singapore Pte Ltd. 2025

S. Wang, M. Gribskov, *Practical Bioinformatics*,
https://doi.org/10.1007/978-981-96-7949-2_6

sequence, we can infer that the structure and function of the novel gene/protein are similar to the known gene/protein in the database because we know that they are descended from the same gene/protein. Generally, we identify many potentially matching genes/proteins from the database search. Greater similarity of a query and subject sequence indicates higher similarity in function and structure. The similarity of the query and subject sequences is measured from their alignment (letter-by-letter sequence matching) and is affected by the type of sequence (nucleotide or protein), the scoring system, and the length of the sequences. This requires a statistical approach to determine whether the match between the query and subject sequence is significant.

In particular, we are interested in whether the observed similarity of the query and subject sequence is significantly greater than we would see between unrelated sequences of the same length and composition. The rationale for this approach was described in in Chap. 5; if we are confident that the query and subject are not unrelated (i.e., we can reject the null hypothesis), we can conclude they are homologous. This analysis depends on the statistics of the extreme value distribution introduced in Chap. 5.

Performing a complete alignment of a query sequence and every sequence in the database is extremely time-consuming. It is a needle-in-a-haystack problem; we want to find a handful of highly similar sequences that are buried among millions of unrelated sequences. Since almost all the sequences are, in fact, unrelated, what we need is an extremely fast way to identify and exclude the unrelated sequences. This is where the extreme value distribution comes in. As with coin tossing, we can evaluate whether the alignment is random by looking at the longest runs of matching letters, although the analysis is made more complicated by the four-letter DNA sequence alphabet and the 20-letter protein sequence alphabet. This works because aligned sequences comprise matching and not-matching positions which makes them behave much like series of heads and tails.

The most widely used sequence database searching program, *BLAST*, relies on the strategy described above. The steps in BLAST are:

Step 1: Index the database so that short fixed-length words can be rapidly looked up. This index is saved, so this step only needs to be done once. When we use public servers such as those at NCBI or UniProt, this index is premade, so we don't see it.

Step 2: Index the query sequence in the same way.

Step 3: Compare the query index to the database index and identify sequences that have at least two matching words and that require no gaps for the matching words to be aligned.

Step 4: Extend the initial word matches to the longest segment that matches without gaps. BLAST calls these hits or high-scoring pairs (HSPs). The highest scoring HSP between two sequences is the maximal segment pair (MSP). The lengths of the HSPs and MSPs follow an extreme value distribution.

Step 5: Test whether the MSP is large enough to be significant (not random).

Step 6: Perform a complete local alignment (with gaps) of sequences that pass the test from the previous step. While it cannot be theoretically proven, we know, empirically, that local alignment scores also follow an extreme value distribution. This is not surprising since the best alignment is the maximum-scoring alignment selected from a huge number of possible alignments.

The highest matching scores are reported along with the expected number of unrelated sequences that would have an equal or higher score. This latter statistic is called the *Expectation* or *E-value*. The E-value is basically a *P*-value that has been corrected for multiple testing. The E-value is related to the number of false positives we expect to see.

We usually use E-values to determine whether sequences are similar enough to be considered homologous because the E-value tells us the number of unrelated sequences that would be likely to score as high as our query and subject sequences. An E-value of 10^{-100} indicates that it is extremely unlikely that alignments of unrelated sequences would ever score as well, while $E = 1$ suggests that you would see at least one match at this level even for unrelated sequences. Note that a score of $E = 1$ does not indicate that the query and subject are conclusively unrelated; it only indicates that they might be. This is because the extreme value distribution only tells us about unrelated (random) sequences—it tells us nothing about the distribution of scores for related (homologous) sequences.

The E-value is determined by the following equation, frequently called the Karlin-Altschul equation (Eq. 3.1)

$$E(\text{Score} \geq S) = Kmne^{-\lambda S} \tag{6.1}$$

where m and n are the sequence lengths (query length and the sum of the lengths of the database sequences). K and λ are constants that depend on the scoring system and the frequencies of the letters (bases or residues) in the sequences. By normalizing the comparison scoring system used to calculate the score S, we can calculate a modified score, S', that includes K (Eq. 6.2); it is easy to see that the E-value depends only on the size of the search space (mn) and the score, S' (Eq. 6.3).

$$\text{if } S' = \frac{\lambda S - \ln K}{\ln 2} \tag{6.2}$$

$$\text{then } E(\text{Score} \geq S') = mn2^{-S'} \tag{6.3}$$

When scaled in this way, the scores are called *bit* scores; this is the raw score reported in the BLAST results in addition to the E-value. The search space, mn, is the product of the length of the query and the lengths of the sequences in the database.

In performing a database search, we test the significance of millions of sequence comparisons. The number of HSPs with $Score \geq S$ in random sequences is described by a Poisson distribution. The probability of finding at least one HSP with $Score \geq S$ is therefore (Eq. 6.4)

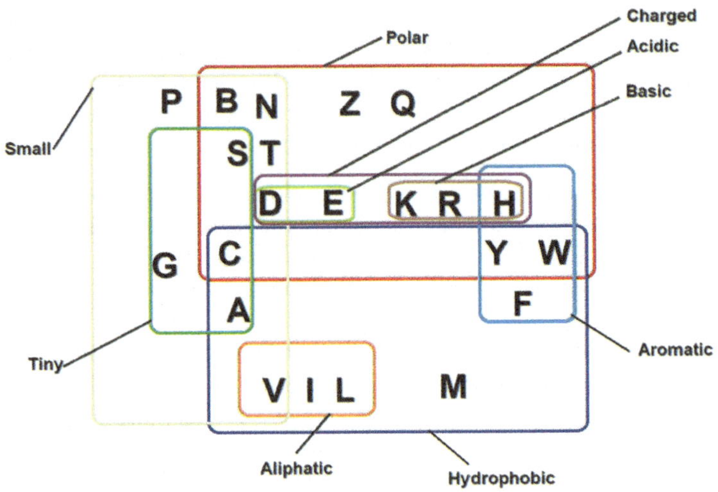

Fig. 6.1 Amino Acid Similarity. A Venn diagram showing the complex pattern of similarity between the 20 amino acid residues

$$P(\text{Score} \geq S) = 1 - e^{-E} \tag{6.4}$$

One of the reasons for using E-values rather than *P*-values is that when E is close to zero, it is difficult to see differences in the *P*-values, even when the E-values are clearly different. For instance, for E-values of 5.0, 10.0, and 20.0, the *P*-values are 0.993, 0.99995, and 0.999999998, respectively. The E-value also directly tells you the number of false positives you expect with a certain score.

The four bases in nucleotide sequences are roughly comparable, and usually no distinction is made between the different mutational changes, for instance, A → T generally has the same score as A → G or A → C. The situation is more complicated for proteins. The 20 amino acid residues have a complex pattern of chemical similarity (Fig. 6.1). For protein sequence comparisons, the best scoring tables are based on empirical tabulations of the frequency of amino acid exchanges in identical positions in known homologous proteins. One widely used scoring system, and the default for the BLAST programs, is called Blosum62 (Fig. 6.2).

Because of the difference in the number of letters in nucleotide sequences, which have four bases, and protein sequences, which have 20 amino acid residues, it is much more likely that random nucleotide sequences will match than that protein sequences will match. For example, the probability of randomly seeing the sequence AAAA in a nucleotide sequence is $0.25^4 = 3.91 \times 10^{-3}$, whereas the probability of randomly seeing a rare amino acid residue (W) or a common amino acid residue (L) four times is $0.012^4 = 2.1 \times 10^{-8}$ or $0.096^4 = 8.5 \times 10^{-5}$, respectively. Since our test for homology is always a comparison to how likely we are to see an equal or higher score in random sequences, it is easier to conclude that protein sequences are

	A	R	N	D	C	Q	E	G	H	I	L	K	M	F	P	S	T	W	Y	V	B	Z	X	*
A	4	-1	-2	-2	0	-1	-1	0	-2	-1	-1	-1	-1	-2	-1	1	0	-3	-2	0	-2	-1	0	-4
R	-1	5	0	-2	-3	1	0	-1	0	-3	-2	2	-1	-3	-2	-1	-1	-3	-2	-3	-1	0	-1	-4
N	-2	0	6	1	-3	0	0	0	1	-3	-3	0	-2	-3	-2	1	0	-4	-2	-3	3	0	-1	-4
D	-2	-2	1	6	-3	0	2	-1	-1	-3	-4	-1	-3	-3	-1	0	-1	-4	-3	-3	4	1	-1	-4
C	0	-3	-3	-3	9	-3	-4	-3	-3	-1	-1	-3	-1	-2	-3	-1	-1	-2	-2	-1	-3	-3	-2	-4
Q	-1	1	0	0	-3	5	2	-2	0	-3	-2	1	0	-3	-1	0	-1	-2	-1	-2	0	3	-1	-4
E	-1	0	0	2	-4	2	5	-2	0	-3	-3	1	-2	-3	-1	0	-1	-3	-2	-2	1	4	-1	-4
G	0	-2	0	-1	-3	-2	-2	6	-2	-4	-4	-2	-3	-3	-2	0	-2	-2	-3	-3	-1	-2	-1	-4
H	-2	0	1	-1	-3	0	0	-2	8	-3	-3	-1	-2	-1	-2	-1	-2	-2	2	-3	0	0	-1	-4
I	-1	-3	-3	-3	-1	-3	-3	-4	-3	4	2	-3	1	0	-3	-2	-1	-3	-1	3	-3	-3	-1	-4
L	-1	-2	-3	-4	-1	-2	-3	-4	-3	2	4	-2	2	0	-3	-2	-1	-2	-1	1	-4	-3	-1	-4
K	-1	2	0	-1	-3	1	1	-2	-1	-3	-2	5	-1	-3	-1	0	-1	-3	-2	-2	0	1	-1	-4
M	-1	-1	-2	-3	-1	0	-2	-3	-2	1	2	-1	5	0	-2	-1	-1	-1	-1	1	-3	-1	-1	-4
F	-2	-3	-3	-3	-2	-3	-3	-3	-1	0	0	-3	0	6	-4	-2	-2	1	3	-1	-3	-3	-1	-4
P	-1	-2	-2	-1	-3	-1	-1	-2	-2	-3	-3	-1	-2	-4	7	-1	-1	-4	-3	-2	-2	-1	-2	-4
S	1	-1	1	0	-1	0	0	0	-1	-2	-2	0	-1	-2	-1	4	1	-3	-2	-2	0	0	0	-4
T	0	-1	0	-1	-1	-1	-1	-2	-2	-1	-1	-1	-1	-2	-1	1	5	-2	-2	0	-1	-1	0	-4
W	-3	-3	-4	-4	-2	-2	-3	-2	-2	-3	-2	-3	-1	1	-4	-3	-2	11	2	-3	-4	-3	-2	-4
Y	-2	-2	-2	-3	-2	-1	-2	-3	2	-1	-1	-2	-1	3	-3	-2	-2	2	7	-1	-3	-2	-1	-4
V	0	-3	-3	-3	-1	-2	-2	-3	-3	3	1	-2	1	-1	-2	-2	0	-3	-1	4	-3	-2	-1	-4
B	-2	-1	3	4	-3	0	1	-1	0	-3	-4	0	-3	-3	-2	0	-1	-4	-3	-3	4	1	-1	-4
Z	-1	0	0	1	-3	3	4	-2	0	-3	-3	1	-1	-3	-1	0	-1	-3	-2	-2	1	4	-1	-4
X	0	-1	-1	-1	-2	-1	-1	-1	-1	-1	-1	-1	-1	-1	-2	0	0	-2	-1	-1	-1	-1	-1	-4
*****	-4	-4	-4	-4	-4	-4	-4	-4	-4	-4	-4	-4	-4	-4	-4	-4	-4	-4	-4	-4	-4	-4	-4	1

Fig. 6.2 Blosum62 amino acid scoring matrix. Values are given in 1/2 bit units. Positive values, shown in red, indicate amino acid exchanges that are more common in homologous proteins than in random protein sequences

homologous than the corresponding nucleotide sequences. One of the main considerations in identifying related sequences using a BLAST search is therefore the type of query sequence. Nucleotide sequences are more prone to random matches than protein sequences because they use only a 4-letter alphabet, while protein sequences use a 20-letter alphabet. One of the goals of this chapter is to investigate the relative power of searching a nucleotide database with a nucleotide or protein query compared to searching a protein database with a protein query.

For flexibility, BLAST accommodates both protein and nucleotide databases and query sequence, resulting in the four primary BLAST programs.

- blastn—nucleotide query searching a nucleotide database
- blastp—protein query searching a protein database

- blastx—nucleotide query translated in all six reading frames searching a protein database
- tblastn—protein query searching a nucleotide database translated in all six reading frames

The four programs are used in somewhat different situations. Blastn and blastp are used to directly find homologs of a nucleotide or protein sequence query, respectively. However, because proteins often show more significant matches than nucleotide searches, blastx is more informative for nucleotide queries, especially queries that represent newly sequenced genomes. Tblastn is much less often than the other programs. The most obvious use is when you have segments of DNA from different species that you believe encode a gene, but the encoded protein sequences are not available.

6.3 Before Lab Questions

Using the introductory material and/or online resources, define the following:

- HSP
- MSP
- E-value

Answer the following questions:

- What type of queries and databases are used by BlastN, BlastP, and BlastX?
- Why does the presence of homology allow us to conclude that a query sequence has a similar structure and function to a subject sequence?

6.4 Resources and Materials

- NCBI/BLAST website: https://blast.ncbi.nlm.nih.gov/Blast.cgi
- DNA sequence of colistin resistance gene (mcr1_dna.ecoli.fa, supplemental file 2)
- Protein sequence of colistin resistance protein (mcr1_protein.ecoli.fa, supplemental file 3)

6.5 Activity 1: BlastN

Use a web browser to go to the URL for BLAST listed in *Resources and Materials*. The four primary blast programs are shown in the middle of page: nucleotide BLAST (BlastN), protein BLAST (BlastP), blastx, and tblastn (Fig. 6.3).

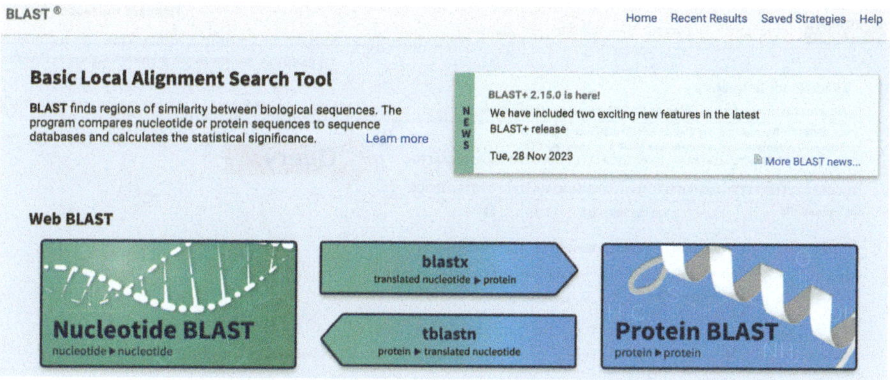

Fig. 6.3 BLAST homepage (https://blast.ncbi.nlm.nih.gov/Blast.cgi). Only the top portion is shown

Step 1: Select *Nucleotide BLAST.*

Step 2: A new page will pop up. Copy and paste the MCR-1 DNA sequence into the box marked *Enter accession number(s), gi(s), or FASTA sequence(s)* (Fig. 6.4). For long sequences, you can limit the search to a segment of the query. In this case, we want to use the whole sequence, so leave *From* and *To* blank. The *Title* should be filled in automatically based on the query, but you can change it if you want.

Step 3: In the section marked *Choose Search Set,* select the default option *Standard databases (nr* etc.*)*, and in the pulldown menu Database, please choose *RefSeq Genome Databse (refseq_genomes)* as the database. We use RefSeq because the search is much faster than if we use the default *Nucleotide collection (nr/nt).*

Step 4: To further reduce the time required for the search, we will restrict the search to a specific taxonomic group, the phylum Firmicutes (more correctly known as *Bacillota*). In the pulldown menu labelled *Enter organism name or id*, enter *firmacutes or firmicutes (both spellings exist in NCBI database)*, the program will provide you a selection menu, choose any option labelled *taxid*: *1239* (1239 is the numeric index of the Firmicutes phylum in the NCBI Taxonomy database).

This will limit the search to only Firmicutes bacteria, reducing the number of database sequences to search and allowing the search to be finished during class.

Step 5: In the section marked *Program* Selection, select *Somewhat similar sequences (blastn)*, instead of the default *Highly similar sequences (megablast)*. The megablast option is faster but much less sensitive.

Step 6 (optional): An additional section labeled +*Algorithm parameters* on a blue bar allows you to adjust the amount of output. If you expect a larger number of matches, you may want to increase *Max target sequences* to 1000 or 5000. You may also want to limit the results to only the most significant matches by reducing *Expect threshold* to 1e-5 or 1e-20. Do not change other parameters unless you know exactly what you are doing because it my negatively affect the time required for the search or the accuracy of the result (Fig. 6.5).

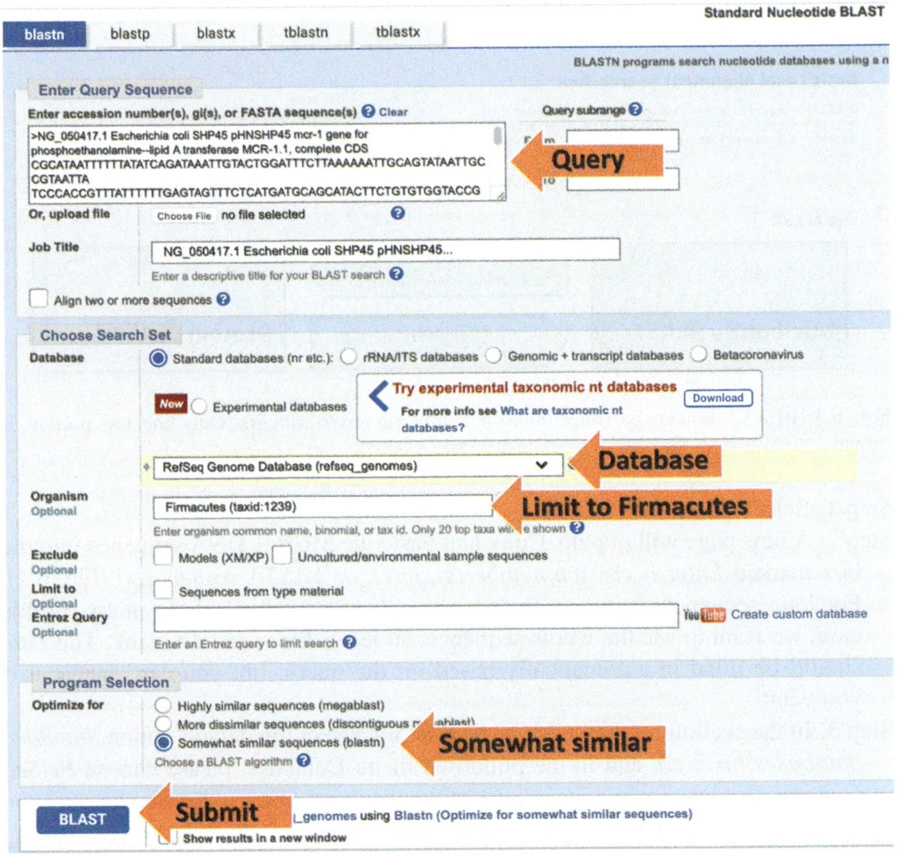

Fig. 6.4 BlastN search interface (https://blast.ncbi.nlm.nih.gov/Blast.cgi?PROGRAM=blastn&PAGE_TYPE=BlastSearch&LINK_LOC=blasthome)

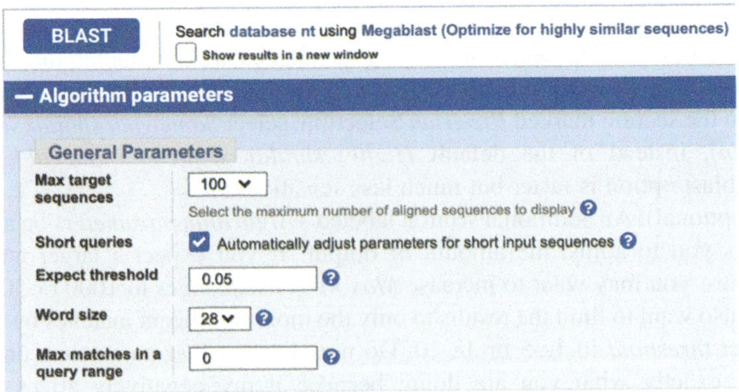

Fig. 6.5 BLAST algorithm parameters

Step 7: Click the blue button *BLAST* to start the search. The search will take several minutes to complete. You will see the results in a new page.

Notebook
Only a few sequences should match this gene, and those very poorly. Look at the E-values and lengths of the matches. Are these genes homologs of the query?

6.6 Activity 2: BlastP

This activity is similar to Activity 1, except you will search a protein database with a protein sequence. You should see a very big difference in the number of homologs you can detect. To make the comparison easier, choose the same limitations to the target database as you used with BlastN.

Step 1: This time, click on the "Protein BLAST." The submission is very similar to BlastN.
Step 2: A new page will pop up, and copy the MCR-1 protein sequence into the query box.
Step 3: Database, please choose "Reference Proteins (refseq-protein)" as the database.
Step 4: Organism, please add term "firmacutes (taxid: 1239)" or "firmicutes (taxid:1239)". This will limit the search to only Firmicutes bacteria.
Step 5: Then click the blue button "BLAST" to start the search.

Notebook
The search will take several minutes to complete, then you will see the results. Record the most significant results in your notebook. Is BlastP more sensitive than BlastN?

6.7 Activity 3: BlastX

BlastX compares a nucleotide query to a protein sequence database by translating the query in all six possible reading frames (three forward and three backward). This gives the search close to the same sensitivity as a search with a protein query. In eukaryotes, genes are often interrupted by *introns* which reduces the ability of BlastX to identify homologs a little because the protein coding sequence may be in different reading frames. In this activity, we are looking at prokaryotes so we don't have to worry about splicing.

Step 1: Click on the "Blastx." Again, this submission is very similar to BlastN.

Step 2: A new page will pop up, and copy the MCR-1 DNA sequence into the query box.

Step 3: Database, please choose "Reference Proteins (refseq-protein)" as the database.

Step 4: Organism, please add the term "firmicutes (taxid: 1239)". This will limit the search to only Firmicute bacteria, so ensuring the search can be finished in the class period.

Step 5: Then click the blue button "BLAST" to start the search.

Notebook

Again, you will see the results in a few minutes. Summarize the results in your notebook.

6.8 Activity 4: Collect Sequences from Diverse Taxonomic Groups

Step 1: BlastP search. Use your aoi protein sequence, we use the Mcr-1 protein sequence (mcr1_protein_ecoli.fa) to query the non-redundant protein sequence (nr) database using protein Blast.

Step 2: More than one search will be necessary. In order to speed up the searches, we will use BLAST's ability to taxonomically limit the search as shown in Fig. 6.6.

In this example, instead of actually cutting and pasting the sequence, we provide the sequence accession number. In order to require fewer searches, it may be helpful to change some of the default settings by selecting "Algorithm Parameters" at the lower left of the query page. These options are highlighted in Fig. 6.6:

- Increasing "Max target sequences" will direct BLAST to report more matching sequences.
- Decreasing "Expect threshold" will make BLAST more stringent, i.e., it will report fewer very weak matches. An "Expect threshold" of 1e-5 or 1e-10 may be helpful.

Step 3: To identify the taxonomically diverse sequences we are interested in, we will take advantage of one of the additional formats provided for BLAST results: the taxonomy report. The process is described in Fig. 6.7.

Look for sequences that are:

1. Full length (the Mcr-1 protein is 541 residues long. 541 $\pm$ 5% is 514 to 568 residues).
2. The match between the query and subject (database) sequences should cover nearly the whole query sequence.

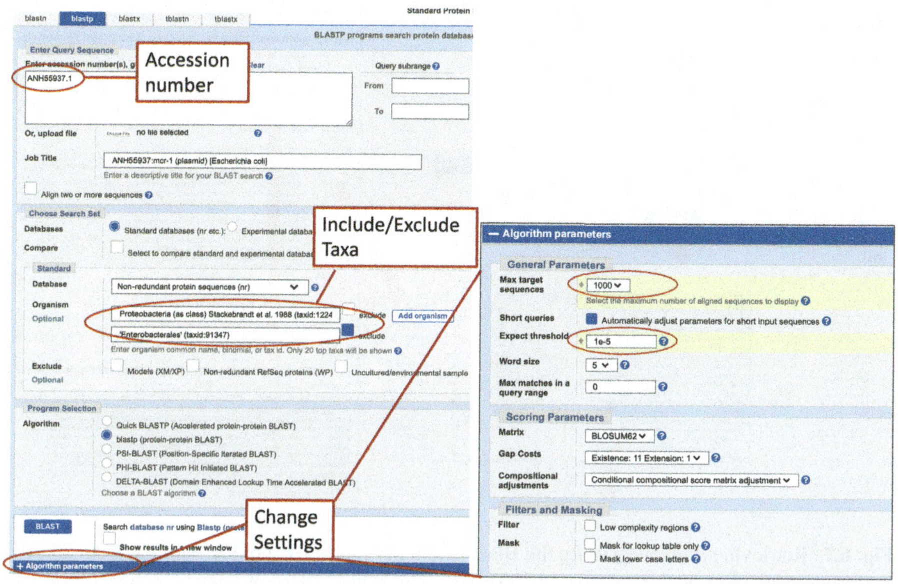

Fig. 6.6 BlastP search (https://blast.ncbi.nlm.nih.gov/Blast.cgi?PROGRAM=blastp&PAGE_TYPE=BlastSearch&LINK_LOC=blasthome). This example shows a search limited to proteobacteria and excluding *Enterobacteriales*

3. An E-value of 1e-10 or less with respect to the *E. coli* Mcr-1 sequence should be sufficient.
4. Completely classified, including genus and species. Do not use unclassified taxa—In future activities we will need the taxonomic information for phylogenetic analysis (Chap. 10).

6.9 Analysis/Notebook

- Why are nucleotide sequences more likely to randomly match than protein sequences?
- The BlastN and BlastX searches are performed with exactly the same query, the MCR-1 nucleotide sequence. Are the results different? Why or why not?
- The results of the BlastP and BlastX searches are nearly identical. Would this be the case if we were looking at a eukaryotic gene that includes exons and introns?

(continued)

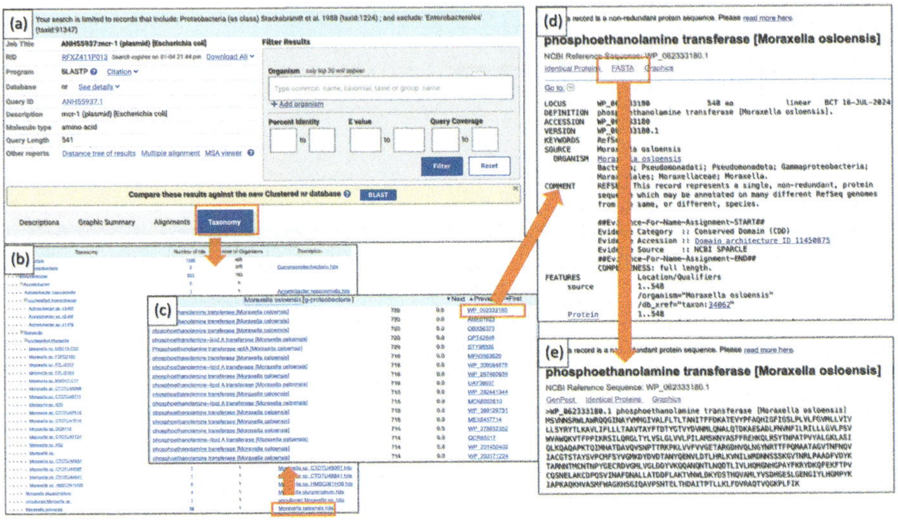

Fig. 6.7 Retrieving sequences using the Blast taxonomy report. Beginning with the main BLAST results page (**a**), click on the taxonomy reports link. This will open a new page with three blocks: the lineage report, the organism report, and the taxonomy report (**b**). The taxonomy report is easiest to use because the various entries can be expanded and compressed using + and −. Clicking on the link in the right column, which will say <species_name> hits, will take you to the entry for this species in the organism report (**c**). In the organism report, you can see both the E-value and a link to the NCBI sequence database (**d**). Once you reach the sequence entry, you can verify the length of the sequence. If you decide to use this sequence, click the link entitled FASTA to get the FastA-formatted sequence (**e**). Cut and paste the entry beginning with and including the > and going through the last sequence letter info in your file of sequences

- When you run BlastN, BlastP, and BlastX, the time it takes to return the results is different. Why?
- For each search, please indicate which hits represent homologous genes/proteins, and explain your reasoning.
- Can a database sequence (hit) that is truly homologous to the query have an E-value >1?

Further Reading

1. The Statistics of Sequence Similarity Scores. https://www.ncbi.nlm.nih.gov/BLAST/tutorial/Altschul-1.html
2. BLAST Glossary. https://www.ncbi.nlm.nih.gov/books/NBK62051/

Chapter 7
Finding Homologous Sequences Using UniProt

From previous activities, students may observe that using protein sequence queries are better able to identify homologs than DNA sequence queries. The Universal Protein Resource (UniProt), which is a collaboration between the European Bioinformatics Institute (EMBL-EBI), the Swiss Institute of Bioinformatics (SIB) and the Protein Information Resource (PIR), is the world's leading high-quality, comprehensive, and freely accessible resource of protein sequence and functional information. UniProt contains information about the genetics, structure, and function in addition to just protein sequences. We will use both text and sequence comparison functions to identify proteins related to your aoi. In addition to finding sequences by sequence comparison (using BLAST), we sometimes want to find sequences with a broad taxonomic distribution. An example of such a case is presented in Chap. 10, where we use phylogenetic trees to look at horizontal gene transfer of antibiotic resistance. In this chapter, we will learn how to select sequences from a broad taxonomic range using UniProt. UniProt also provides an alternative way to find a broadly distributed set of sequences using the UniRef database.

7.1 Objectives

- Select protein sequences related to resistance to your aoi from a broad taxonomic range by comparing to a clustered database.
- Select protein sequences relevant to your aoi using text based searches.
- Prepare sequences for later phylogenetic analysis.

© The Author(s), under exclusive license to Springer Nature Singapore Pte Ltd. 2025

S. Wang, M. Gribskov, *Practical Bioinformatics*,
https://doi.org/10.1007/978-981-96-7949-2_7

7.2 Background

This book focuses on getting clues about the structure and function of a protein based on related sequences. In essence, nature and evolution are an enormous testing laboratory, where thousands of mutations are tested to see which are functional. This is particularly the case for genes that are subject to strong selection, such as antibiotic resistance genes. To take advantage of this natural laboratory, we want a set of sequences with very broad sequence diversity. What we typically get with a BLAST search is a set of sequences that are very close to our query, *i.e.*, not diverse at all. The solution lies in using a database that has already had very similar sequences merged into single entries, the UniProt UniRef50 database. The UniRef databases consist of groups of sequences where every sequence with more than a certain percent identity after aligning is represented by a single sequence. Clustering levels of 100%, 90%, and 50% are available. Using a clustered database decreases the size of the database, making searching faster, and ensures that the subject sequence hits have a minimum difference from each other. UniProt provides a version of BLAST at its website. The interface is somewhat simpler than the one at NCBI.

7.3 Resources and Materials

- UniProt database—https://www.Uniprot.org/
- Protein Sequence of Colistin resistance protein (mcr1_protein.ecoli.fa, supplemental file 3).
- Table 7.1 for search terms.
- Table 7.2 for Highly sequenced bacterial phyla.

7.4 Activity 1: BLAST Using UniProt Database

Step 1: Using the MCR-1 protein sequence, perform a BLAST search at the UniProt database[1] (Fig. 7.1). Change the following parameters from the defaults

- Target Database—Change to UniRef50,
- E-threshold—change to the smallest value 0.0001 (1e-4),
- Hits—increase to the largest value, 1000.

Step 2: After the search is complete, the results show the alignment of the representative sequence from each matching cluster to your query sequence. The diagram

[1] The UniProt Consortium, UniProt: a worldwide hub of protein knowledge, Nucleic Acids Research, Volume 47, Issue D1, 08 January 2019, Pages D506–D515.

Table 7.1 Examples of antibiotic resistance search terms

Search term(s)	Selected gene names	Expected hits (approximate)
"Aminoglycoside nucleotidyltransferase"	aadA, B, E, K	668
"Aminoglycoside 6 adenylyltransferase"	aadE,K	1092
Aminoglycoside N(3)-acetyltransferase	aacC1,2,3, yokD, neo	4564
Erythromycin esterase	ereA,B	4063
"Butirosin biosynthesis protein h"	btrH	329
Undecaprenol kinase	dgkA	988
"Ribosomal rna adenine demethylase"	rsmA, ksgA, ermA,B,G	527 bacteria
Multidrug efflux pump subunit AcrB	acrB,E bepE	4380
Chloramphenicol acetyltransferase	CatA,B,D,3, cml	8325 bacteria
"Vanz family"		6435
Chloramphenicol phosphotransferase	cpt	865 bacteria
"Beta lactamase b"	Blm, cph,blab, ccr	4059
yvqf	yvqf	2659
Streptothricin acetyltransferase	satA, sta	2109 bacteria
Streptogramin A acetyltransferase	vatD	399 bacteria
Fosfomycin	fosA,B,X, abaF	5514 bacteria
Lincosamide nucleotidyltransferase	linB, lnuA,B,G, aadA2	348 bacteria
Ampc "beta lactamase"	ampC	2657 bacteria
Rifampin ADP-ribosyltransferase	arr, arr2,3	537 bacteria
"Irginiamycin b lyase"	vgb	1544 bacteria
Macrolide glycosyltransferase	Mgt, oleD	286
Vancomycin/teicoplanin A-type resistance protein D-alanine—D-alanine ligase	vanA	1091 bacteria
Tunicamycin resistance	tmrB, alg7	411 bacteria
Polymyxin resistance	arnA	2849 bacteria
Penicillin binding protein PBP4	Pbp4, dacA	310 bacteria
Penicillin binding protein PBP2	Pbp2, mecA, mrdA,	181
"Nitroimidazole antibiotic resistance"	nimA,B	815

on the right of the search result (Fig. 7.2) shows the fraction of your query sequences that matches to the cluster. Preferably, the match is close to full length. By holding your mouse pointer on the bar, you can see the hit length. Select sequences that have similar hit lengths and overall length (Length column) as much as possible. Select sequences by clicking the small box to the left of the Cluster ID; try to select sequences from a diverse range of taxa (Organism column). Then click Add (above the blue column headings bar) to add the sequences of the representative sequences to your "shopping basket".

Table 7.2 Highly sequenced bacterial phyla

Phylum	Genomes sequenced	Phylum	Genomes sequenced
Acidobacteriota	2028	Cyanobacterota	3846
Actinomycetota	33994	Desulfobacterota	2143
Bacillota	74157	Myxococcota	1033
Bacillota_A	33495	Patescibacteria	5830
Bacillota_C	1803	Planctomycota	2450
Bacteroidota	29028	Pseudomonadota	174676
Campylobacterota	8436	Spirochaeota	3007
Chloroflexota	3064	Verrucomicrobiota	3678

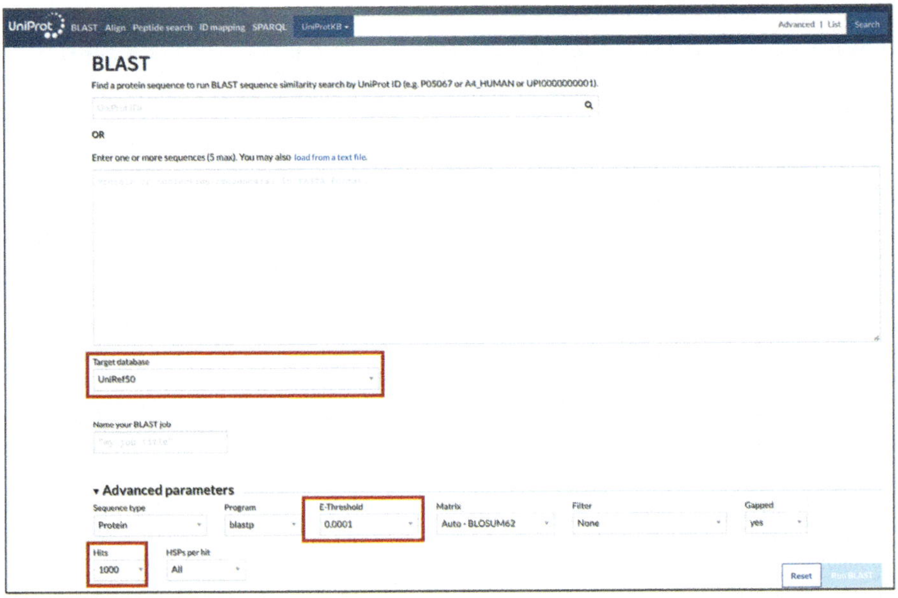

Fig. 7.1 UniProt BLAST Search (https://www.uniprot.org/blast). The parameters that need adjustment are squared

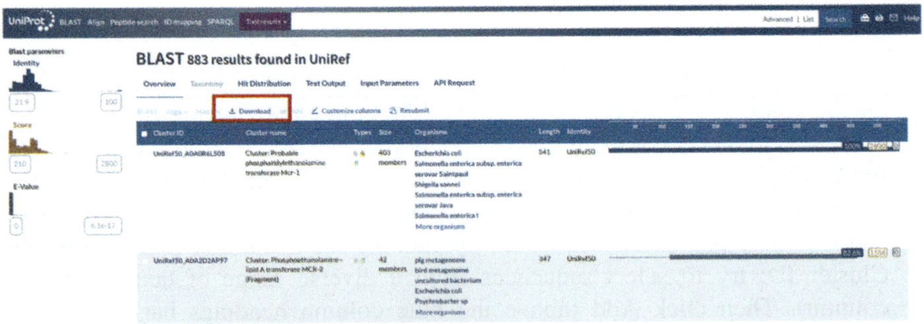

Fig. 7.2 Downloading BLAST results from UniRef

Step 3. When you are finished adding sequences, download the files to your browser by selecting the "shopping basket" in the upper right of the blue main menu bar, and selecting Download. Be sure to select 'Download all,' FASTA (representative), and 'Compressed=No'. In my case, using the chrome web browser, the sequences were automatically saved to a file. Copy this file to your working directory and rename it to somethings sensible (*e.g.*, UniRef50_seq.fa). If your browser displays the file on the screen, save it to a file in the normal way.

Step 4. Inspect your sequences and remove sequences using a text editor and remove sequences that are much longer or shorter than your query (approximately +/- 5%).

Step 5. Rename the header line of each sequence as discussed in the following activity.

7.5 Activity 2: Selecting Sequences from UniProt

To find evidence for horizontal transfer, we want to select sequences for antibiotic resistance proteins from a diverse group of genera. We will use the text searching function of the UniProt database, which contains taxonomic annotations, to do this.

Step 1: Go to https://www.Uniprot.org/UniProt. UniProt uses a shopping basket metaphor to select groups of sequences.

Step 2: Table 7.1 shows some examples of search terms related to various antibiotic resistance proteins/genes. Use the example search terms to learn how the search interface works. Then use search terms related to your own aoi to find relevant information. Start by typing the search term in the query box. For instance, a search term might be *Aminoglycoside 3′-phosphotransferase*. If your search term includes more than a single word, as in this example, place quotation marks around it. This requires the entire phrase to match. You will probably see a very large number of matching sequences—in this case, I see 2991 results (Fig. 7.3).

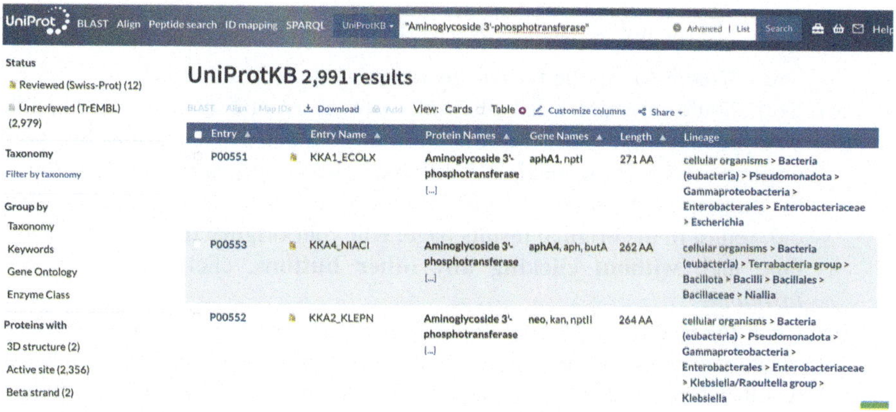

Fig. 7.3 Example of UniProt search

Step 3: The first task is to get an approximate idea of the usual size of the protein. Many proteins contain multiple domains, and for phylogenetic analysis, we must be sure that all the sequences have the same homologous domains. Scan through the list of result, looking at the names and sequence lengths. If I search for *aminoglycoside phosphotransferase*, I see two kinds of results—some proteins are labeled *Aminoglycoside 3′-phosphotransferase* and have lengths around 360–280 residues. Others are labeled *Bifunctional AAC/APH* and have lengths around 480 residues. I could choose either type, but the shorter form is more likely a single domain protein, and appears to be more common, so I choose to work on it.

Step 4: Now I want to select sequences belonging to as many genera of bacteria as possible. It is difficult to do this with the default output because only the genus and species are shown. However, with a few clicks, it is possible to see the higher taxonomic levels for each result.

Before we select sequences, here are the requirements for selecting sequences.

Requirement 1: Select sequences from at least five phyla shown in Table 7.2. Two of the phyla **must be** ***Pseudomonadota*** **and** ***Bacillota***.

Requirement 2: Select sequences from at least 40 genera. To see higher taxonomic levels, click on the Genus/species in the Organism column. The Lineage shows a compressed version of the taxonomy for the entry. For instance, for Escherichia coli, I see Pseudomonadota > Gammaproteobacteria > Enterobacterales > Enterobacteriaceae > Escherichia which represents the Phylum > Class > Order > Family > Genus. It may also be helpful to refer to the GTDB search page (https://gtdb.ecogenomic.org/searches)

Requirement 3: Choose only sequences that are completely classified, including genus and species. Do not use unclassified taxa or taxa listed as *incertae sedis* (Latin for uncertain placement) – we will use these sequence to construct phylogenetic trees and look for horizontal transfer (Chap. 11) and sequences with unknown phylogenetic placement are not useful.

There are two approaches to selecting sequences.

Approach 1: Select using taxonomy view (IMO, the easiest way)

(i) Select *Taxonomy* in the *Group by* menu at the left of the page.
(ii) You should now see a list of results shown as an expandable/collapsible phylogenetic tree. Double-check that the number of results matches your original search. If the number is different, check the search box at the top of the page to make sure that UniProt did not change your search query—if it did, return to the original results page, edit your original query in the search box, and, **without clicking any other buttons**, click *Group by -> taxonomy.*
(iii) Click the triangle to expand the taxa that you want to sample (Fig. 7.4).
(iv) When you reach the desired level, click on the link that shows the number of results (shown immediately to the left of the taxonomic name). For instance,

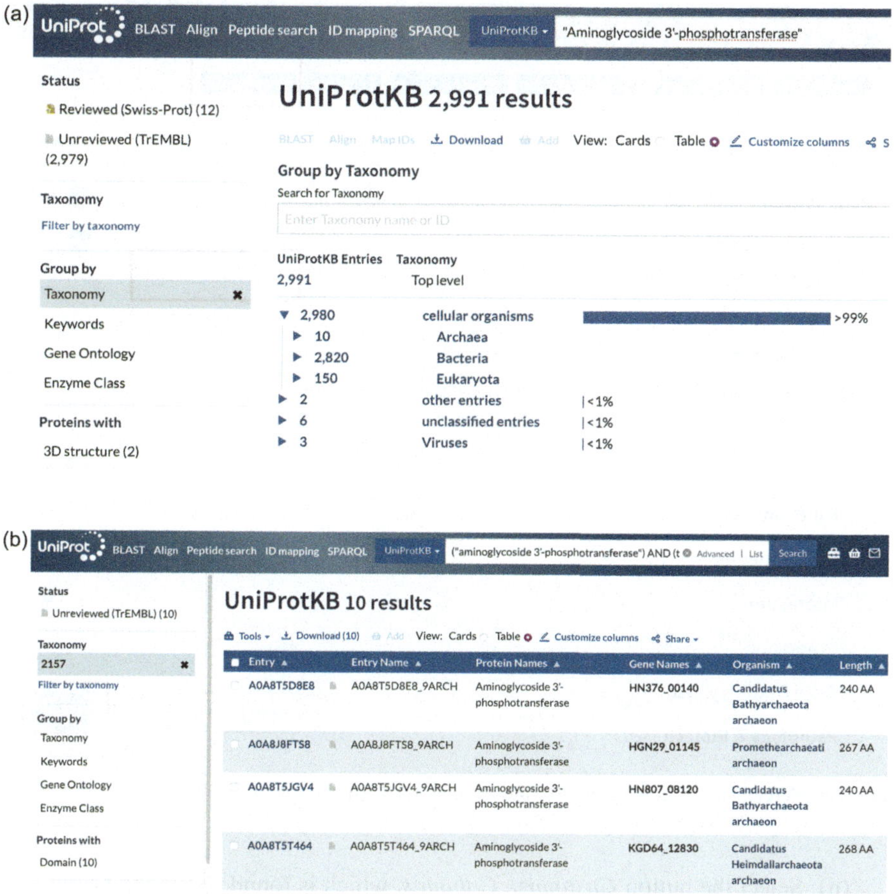

Fig. 7.4 Examples of UniProt search results. (**a**) Screenshot of the search results grouped by Taxonomy. (**b**) Example of Archaea results

in Fig 7.4a, the number for Archaea is 10, and the number for bacteria is 2,820. Clicking on 10 brings us to a page that shows all 10 archaeal aminoglycoside 3'-phosphotransferase sequences (Fig. 7.4b).

(v) Select sequences by clicking on the box to the left of the Entry column and then clicking on Add in the menu bar just above the list of sequence entries to add the sequences to your shopping basket.

Approach 2: Add lineage information and modify the settings

(i) We can modify the settings so that it includes the complete taxonomic lineage.

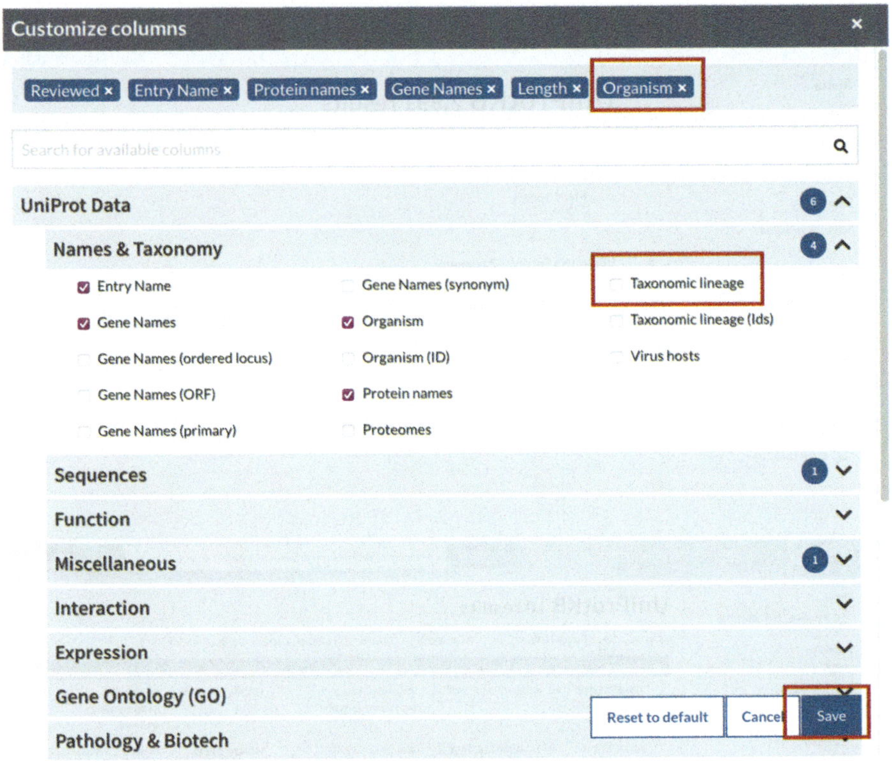

Fig. 7.5 Screenshot of how to modify the settings for UniProt search results

(ii) Select the button *Customize Columns*, which is found above the bright blue bar just above the list of matching sequences.

(iii) In the menu item "Names and Taxonomy", deselect the *Organism* checkbox and select *Taxonomic lineage* instead (Fig. 7.5).

(iv) Select *Save*, a blue button found in the upper right corner of the *Customize results table* page, to return to the list of sequences matching your keyword.

Whichever approach you take:

(i) Select the sequences you want by checking the box at the left of each result.

(ii) Select *Add to basket,* a light blue button found in the menubar just above the results list.

(iii) When you do this, you should see that the number of sequences shown in red in your *Basket* button, found at the extreme top right of the page, goes up by one.

After selecting your sequences and adding them to the "basket," click on the basket to open it and select "Download." Get the sequences in uncompressed Fasta (canonical) format.

Step 5: Edit the sequence's header line. You will need to edit your sequence files to make the sequence names more informative.

Each sequence in your file begins on a line beginning with the ">" character. The first line contains information about the sequence, including its ID, and the following lines, up to the next line beginning with ">," are the amino acid sequence (one-letter code). You should have 40 or more sequences, all in one file. One of the sequences I downloaded has the following title line:

> Original sequence ID lines (UniProt)
> \> tr|A4SUQ2|A4SUQ2_AERS4 Putative sulfatase OS = Aeromonas salmonicida (strain A449).

The sequence ID is shown in the text following the ">," up to the first space - tr|A4SUQ2|A4SUQ2_AERS4. Multiple sequence alignment (MSA) programs and the trees based on MSAs (discussed in the Chaps. 8 and 10) will use this ID as a label. Unfortunately, this label is not very informative, so it is helpful to rename the sequences. Generally, anything after the first space is ignored, so we can insert a modified ID between the > and the current ID. An informative label for the MSA and tree construction activities would include some taxonomic information, I suggest using an abbreviation of the taxonomic class, genus, and species. The names you use have to be unique within the file, and you should separate multiple terms with an underline. *Aeromonas* is in the class *Gammaproteobacteria,* so I might use a name like gprot_ae_salmonica, producing.

> Modified sequence ID lines (UniProt)
> \> gprot_ae_salmonica tr|A4SUQ2|A4SUQ2_AERS4 Putative sulfatase OS = Aeromonas salmonicida (strain A449).

> **Notebook**
> Record the original IDs (accession numbers) and the modified IDs in your notebook. Provide a complete table of the protein sequence identifiers, sequence lengths, and descriptions selected for your tree. Include the Phylum, class, order, family, genus, and species. If you want a complete description of your selected sequences you can do the following:
>
> - In your Basket, select Full view—this will display a results table of the selected sequences.
> - At the top of the page, select download.
> - Choose Format = Tab-separated.

(continued)

- Choose Uncompressed.
- Select Go.

The downloaded information will be exactly what you see on the screen. You can change the information shown using the Customize Columns button as in Sect. 7.5. Summarize in a table the number of selected sequences on the phylum, class, and order levels. If necessary, update information on gene names in your notebook (including names newly identified while selecting sequences).

7.6 Analysis/Notebook

- Summarize in a table, including the number of selected sequences on the phylum, class, and order levels.
- Use online resources to find out about the mechanism of antibiotic resistance in your chosen family and include this in your Notebook. General resources such as Wikipedia and Google can be used. Cite your sources, using primary sources (scientific literature) where possible. And answer the following questions:

  1. What are the gene names and functional synonyms for the proteins in your aoi family?
  2. What is the approximate length of the sequences in the aoi family?
  3. In how broad a taxonomic ranges is the gene/protein for your aoi found?
  4. What antibiotics does the protein family provide resistance to?
  5. How do the proteins cause resistance?
  6. Is the structure of a protein in this family known?
  7. How widespread is the resistance?
  8. Is resistance a serious medical problem? How many patients are affected per year?

Further Reading

1. Genome Taxonomy Database, Parks DH, Chuvochina M, Waite DW, Rinke C, Skarshewski A, Chaumeil PA, Hugenholtz P. A standardized bacterial taxonomy based on genome phylogeny substantially revises the tree of life. Nat Biotechnol. 2018;36(10):996–1004. https://doi.org/10.1038/nbt.4229.

Chapter 8
Multiple Sequence Alignment

We have introduced the concept, statistical background, and databases used for sequence alignment. In addition to pairwise sequence alignment, multiple sequence alignments (MSA) provide a lot of information that can be used to predict the structure and function of nucleotide or protein sequence families. The structure and function of proteins constrain the evolution of the sequences. Regions that are important for protein structure and function accumulate changes more slowly because selection tends to remove individuals with nonfunctional proteins from the population (assuming a protein that affects fitness). MSA therefore gives powerful clues to where the important functional regions of the protein lie. We have already learned how to select sequences using Blast and UniProt (Chap. 7), we will use these selected sequences to construct a multiple sequence alignment in this chapter.

8.1 Objectives

- Construct a MSA of diverse members of an antibiotic resistance gene family.
- Identify conserved (evolutionarily constrained) regions from MSA.

8.2 Background

Multiple sequence alignment (MSA) has been a very active topic of bioinformatics almost from the beginning. Exactly solving the MSA problem is extremely difficult, so all major methods rely on an approximation called *progressive alignment*. In this approach, we first determine the distance between each pair of sequences. We use these distances to construct a tree in which the most similar sequences lie close

© The Author(s), under exclusive license to Springer Nature Singapore Pte Ltd. 2025
S. Wang, M. Gribskov, *Practical Bioinformatics*,
https://doi.org/10.1007/978-981-96-7949-2_8

together. This tree, called the *guide tree,*[1] shows which sequences are most similar and specifies the order in which sequences are added to the MSA. Beginning with the most similar sequences, we align them to each other. Because these sequences are very similar, they have the lowest chance of having incorrect regions in their alignments. After aligning these sequences, their alignment is frozen, and we then align the MSA containing the two sequences to the next closest sequence. By frozen, we mean that the alignment between the first two sequences is not allowed to change, although insertions and deletions with respect to the new sequence can be added. This process is continued, always aligning the closest sequences first, until all sequences have been added to the MSA. The progressive alignment approach is fast and can be used to align thousands of sequences, if necessary.

The drawback to the basic progressive alignment strategy has to do with the freezing of the current MSA before adding each new sequence. If the alignments could be guaranteed to be correct at each step, this would not be a problem. Unfortunately, this is not the case, and errors that occur in the early steps are then "frozen in" and carried through all the way to the final alignment. In the last 15 years, a new generation of MSA programs has addressed this issue by adding an additional step in which subsets of sequences are removed from the MSA, unfrozen, and realigned to the MSA. In many cases, while a pairwise alignment might contain some errors, the alignment of one or a few sequences to the entire group is more likely to be correct. The MSA program that we will use in this lab, Clustal Omega, uses this enhanced progressive alignment strategy.

8.3 Before Lab Questions

- What is meant by a progressive alignment?
- Why are progressive alignments not necessarily optimal?

8.4 Resources and Materials

- UniProt Blast search—http://www.Uniprot.org/blast/
- Clustal Omega website—https://www.ebi.ac.uk/jdispatcher/msa/clustalo
- MView—https://www.ebi.ac.uk/jdispatcher/msa/mview
- WebLogo 3—http://weblogo.threeplusone.com/create.cgi
- Mcr-1 protein sequence, mcr1_protein_ecoli.fa (supplemental file 3).
- Your aoi protein sequence.

[1]Guide trees are constructed using distance-based phylogenetics tree construction methods. Chapter 10 has more information on constructing trees.

8.5 Activity 1: Editing Sequence Names

For the next activity, we need a set of sequences with a broad taxonomic range and meaningful names. Chapters 6 and 7 explain how to select groups of sequences from NCBI and UniProt, respectively. You should have saved these sequences as a single FastA formatted file. As mentioned in Chap. 7, the identifiers (names) of the sequences we retrieve are not very informative—they are just arbitrary IDs without any taxonomic information. To make it easier to interpret the results of this lab and the lab in Chap. 10, we need to rename the sequences. If you already edited your sequence names, as was suggested in Chap. 7, you can skip this activity and proceed to the next activity. If not, please follow instructions in Chap. 7 Activity 2 which explains in detail how to rename the FastA formatted sequences. Briefly, use a text editor that does not insert formatting, such as Notepad (Windows) or TextEdit or TextWrangler (MacOS), to edit the IDs as described below. Do not use Microsoft Word because it inserts invisible characters and line breaks that corrupt the format of the sequence file, making it unusable. It is most convenient to store your edited sequences in a single FastA formatted file.

Please consider the following requirements:

- For the following exercise, you will need 50–100 sequences; It's a good idea to have more because you may have to remove some sequences because they are incomplete, non-homologous, or otherwise incorrect.
- You also need to have sequences from a broad phylogenetic range, so try to sample many different groups rather than having sequences from a single class or order. In Chap. 7, we suggested obtaining sequences from at least five bacterial phyla and 40 genera.
- Your sequences should all be about the same length (approximately +/− 5%). Much longer or shorter sequences are probably due to gene modeling errors or belong to a different sequence family. Remove such sequences while you are editing the sequence names.

In Chaps. 6 and 7, you selected an aoi and a family of antibiotic resistance proteins and saved them as a single text file in *FastA format*. In FastA format the first line contains information about the sequence, including its ID, and the following lines are the amino acid sequence (one-letter code). You should have 40–50 sequences, all in one file. One of the sequences I downloaded has the following title lines when downloaded from GenBank and UniProt, respectively.

Original sequence ID lines (GenBank and UniProt)

>ABO92624.1 putative sulfatase (plasmid) [Aeromonas salmonicida subsp. salmonicida A449]

>tr|A4SUQ2|A4SUQ2_AERS4 Putative sulfatase OS = Aeromonas salmonicida (strain A449).

The sequence ID is shown in the text following the ">," up to the first space; The ID for the Genbank and Uniprot sequences are ABO92624.1 and tr|A4SUQ2|A4SUQ2_AERS4, respectively. Multiple alignment programs (and the trees based on the MSA) will use this ID as a label. Unfortunately, these labels are not very informative, so it is helpful to rename the sequences. Generally, anything after the first space is ignored, so we can insert a modified ID between ">" and the current ID. An informative label for the MSA and tree construction activities would include some taxonomic information, I suggest using an abbreviation of the taxonomic class, genus, and species. The names you use have to be unique within the file, and you should separate multiple terms with an underline. *Aeromonas* is in the class *Gammaproteobacteria,* so I might use a name like gprot_ae_salmonica, producing.

Modified sequence ID lines (GenBank and UniProt)

>gprot_ae_salmonica ABO92624.1 putative sulfatase (plasmid) [Aeromonas salmonicida subsp. salmonicida A449]

>gprot_ae_salmonica tr|A4SUQ2|A4SUQ2_AERS4 Putative sulfatase OS = Aeromonas salmonicida (strain A449).

Notebook

Record the original IDs (accession numbers) and the modified IDs in your notebook.

8.6 Activity 2: Constructing a MSA

We will use the Clustal Omega program to make a MSA. Clustal Omega[2] is a modern progressive alignment program that improves its initial alignment by iteratively removing and realigning sets of sequences. These iterations include reconstructing the guide tree and realigning the sequences to an overall model of the MSA using a hidden Markov model (HMM).

Step 1: Go to the Clustal Omega website at EBI and copy and paste the sequences you prepared in Activity 1 into the query box or upload the FastA file containing your sequences (Fig. 8.1).

Step 2: Modify the settings. Click on the "More options" link on the submission page and change the value of the *"NUMBER of COMBINED ITERATIONS" to 5, "MAX GUIDE TREE" to 2, and "MAX HMM ITERATIONS" to 3.* Usually

[2] Madeira F, Madhusoodanan N, Lee J, et al. The EMBL-EBI Job Dispatcher sequence analysis tools framework in 2024. Nucleic Acids Research. 2024 July;52(W1):W521–W525.

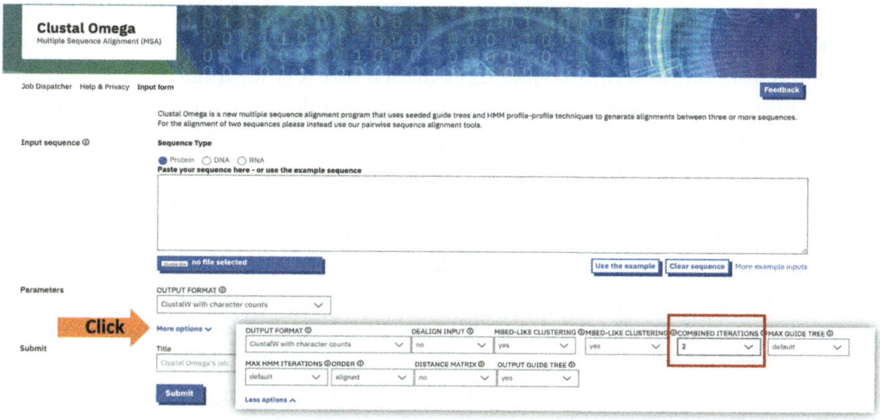

Fig. 8.1 Clustal omega homepage (https://www.ebi.ac.uk/jdispatcher/msa/clustalo). Red box shows optional parameters to change

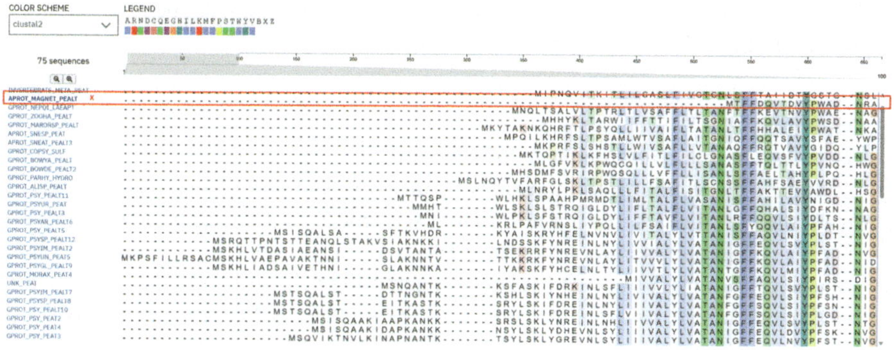

Fig. 8.2 Clustal Omega MSA for colistin sequences (partial results)

you won't see much change with more than 3 iterations. More iterations require more time, but the program runs quickly so feel free to try more.

After submitting the alignment request using the blue Submit button, you will see an intermediate page that shows the job has been submitted, queued, and then is running. For our colistin example with 75 sequences the job required less than a minute to be completed.

Step 3: After the result appears (Fig. 8.2), you have multiple options for coloring the alignment. Coloring helps you to easily identify the most conserved regions. Try various color schemes, my personal favorites are *taylor* and *zappo*. Examine the alignment carefully. Conserved regions are seen as columns that have predominantly the same amino acid residue or background color, for example columns corresponding to the sequence LIIVVALYLVAT indicated in Fig. 8.2. If some of the sequences are shorter and do not cover clearly conserved regions such as the one just mentioned, they should be removed. In this example, I would remove the APROT_MAGNET_PEAL, which is marked with a red X in Fig. 8.2. Check both the left hand (amino-terminal) and right-hand (carboxyl-terminal) ends for incomplete sequences. Also remove sequence that match do not seem to actually match with the other sequences – they are probably not homologs. After removing incomplete and unrelated sequence, repeat the alignment

Step 4: A second version of the MSA can be found in the Tool Output tab. This is the same MSA but it is broken up into blocks for easier viewing as text (for instance in a paper). At the very bottom of the display is a button that will color the MSA using Clustal2's default color scheme (making it easier to see conserved regions). More importantly, under each block of sequence, conserved positions are indicated with the symbols *, :, and ., which indicate residues that are present in all sequences, strongly chemically similar, and weakly chemically similar. Amino acid residues are conserved because they play important roles in the structure and function of the protein, and hence mutations at these positions are rejected by the process of selection. To save the alignment, click the *Tool Output* tab and select download. The file is in clustal format. We will use this alignment in Chap. 10. You can also cut and paste the alignment into a file, but this can be problematic if you do not get the entire result.

Notebook

Save the complete alignment in your notebook.

Make a table noting the locations of the most conserved sequence regions using the coordinates of your aoi protein sequences (not the alignment coordinates). This requires comparing the MSA and the original aoi sequence. You can identify the most conserved regions using the *, :, and ., characters added by Clustal Omega, or using the consensus lines added by MView (see Sect. 8.7)

Step 5: Interpreting the results.

In Fig. 8.2, you can easily see that the sequences tend to be variable at the ends. This is normal variability for protein sequence families and should not cause problems in downstream analysis. However, very long extensions usually indicate that two adjacent genes have been fused together and should be removed. If you screened the sequence lengths and removed very long and short sequences when you renamed the sequences from activity 1, you should not have this problem.

ClustalO also produces a phylogenetic tree in both a text and graphical format. Both are shown in the *Phylogenetic Tree* tab. At the top, you will see the tree in text form (called Newick format, Fig. 8.3). This form is intended mainly for use by visualization programs. At the bottom of the *Phylogenetic tree,* there is a visualization of tree. Unfortunately, large trees are often truncated. We'll look at other tree visualizations in Chap. 10.

```
Alignments          Guide Tree          Phylogenetic Tree

(
(
(
(
(
(
gprot_marsp_PEALT:0.25656,
gprot_pse_hydro:0.25877)
:0.01107,
(
(
gprot_copsy_sulf:0.28134,
gprot_alisp_PEALT:0.23551)
:0.00325,
gprot_parhy_hydro:0.25191)
:0.00458)
:0.00422,
(
invertebrate_meta_PEAT:0.25408,
gprot_zooha_PEALT:0.23088)
:0.02043)
:0.00294,
(
```

Fig. 8.3 Part of the phylogenetic tree from ClustalOmega in Newick text format

8.7 Activity 3: Visualizing MSA with MView

The Results Viewers tab gives you several options for visualizing the alignment and phylogenetic tree. The options on this page actually run additional analysis programs using the EBI Job dispatcher. They automatically create new job submission pages with the necessary information from your Clustal Omega job filled into the form in advance. One of these, available via the EBI job dispatcher at EBI is MView.[3]

Step 1: Open the *Results Viewers* tab and select *Send to MView.*

Step 2: You will see a query page that has been automatically filled in with the output of your Clustal Omega run (mine was named *clustalo-I20250105-143401-0008-83594627-p1m*).

[3]Madeira F, Madhusoodanan N, Lee J, et al. The EMBL-EBI Job Dispatcher sequence analysis tools framework in 2024. Nucleic Acids Research. 2024 July;52(W1):W521–W525.

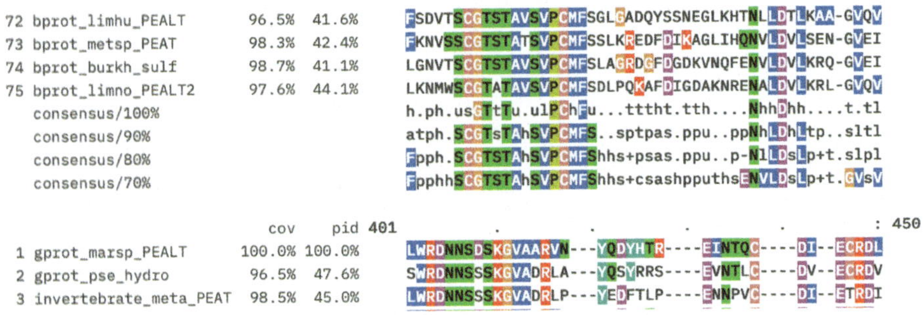

Fig. 8.4 Mview visualization of the colistin MSA (partial result)

Step 3: Select *More options* to see the many parameters that can be controlled. The most important are the *Alignment Width, Color Map,* and *Consensus.* Make sure that *Consensus* is set to *On.* Try various *Alignment Widths* and *Color Maps.* One use of *Alignment Width* is to get a section of interest nicely placed in a block for a manuscript figure. The *Coloring* option controls what positions are shaded. *Identity*, the default, colors only positions where all residues in a column are identical.

Step 4: Add a title if you like, and click *Submit.* The alignment results are exactly the same as the clustal Omega MSA, but you can control how the alignment is broken into blocks and colored (Fig. 8.4). Take particular note of the four lines at the bottom of each block where amino acid residues common to 100%, 90%, 80%, and 70% of the sequences are shown.

8.8 Activity 4: Create a Sequence Logo Using the WebLogo 3 Website

In a sequence logo, the size of the letters represents the conservation of each amino acid residue at each position. The highly conserved positions, at which one residue predominates, are the ones with the biggest letters.

Step 1: Go to the WebLogo 3[4] website (https://weblogo.threeplusone.com/create.cgi).

Step 2: Upload the MSA that you saved from the Clustal Omega program.

Step 3: Almost everything in the sequence logo is customizable: the number of characters per line, coloring, sequence ranges, and logo. Try the various controls to see how they work (Fig. 8.5).

[4]Crooks GE, Hon G, Chandonia JM, Brenner SE WebLogo: A sequence logo generator, Genome Research, 14:1188–1190, (2004).

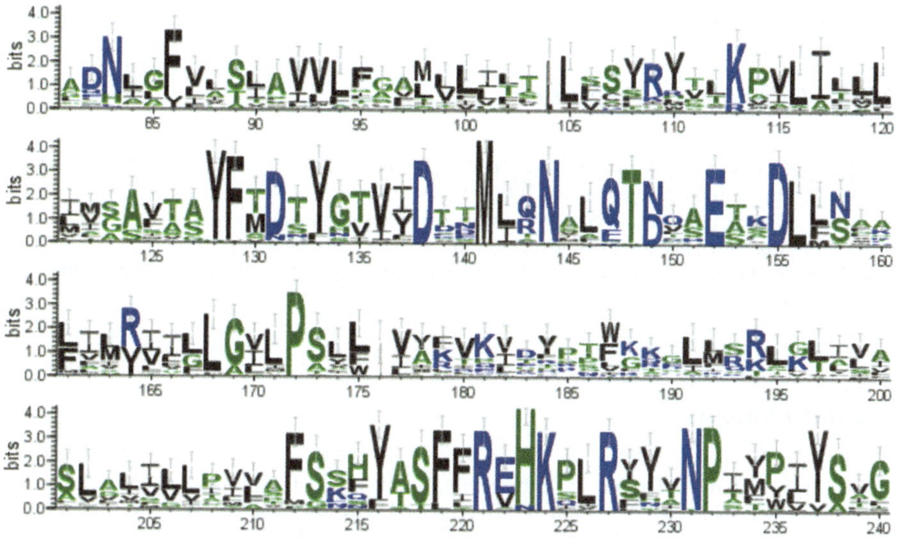

Fig. 8.5 Sequence logo of the colistin MSA (partial results). The region shown corresponds to the region shown in the Mview result. Notice how much more compact the logo is

8.9 Analysis/Notebook

Record the accession numbers of the sequences used in your multiple sequence alignment, as well as the total number of sequences.

- Include a figure showing the multiple sequence alignment, colored by MView.
- Make a table showing the locations of the most conserved sequence regions using the coordinates of your aoi resistance protein sequences (not the coordinates of the alignment). This requires comparing the MSA and the original sequence. Include the conserved sequences.
- Include sequence logos for the 5 longest conserved regions in the table, above.

Answer these questions in your notebook.

1. Why do we remove incomplete or unrelated sequences from the alignment?
2. Why might the sequences not appear in the phylogenetic tree exactly where we expect them to from the known taxonomy.
3. Do any of these conserved regions observed in the alignment or the MView visualization correspond to regions annotated as functionally important in the sequence entry?

Do any of these conserved regions correspond to regions annotated as functionally important in the sequence entry?

Further Reading

Multiple Sequence Alignment

1. Edgar RC, Batzoglou S. Multiple sequence alignment. Curr Opin Struct Biol. 2006;16(3): 368–73. https://doi.org/10.1016/j.sbi.2006.04.004.
2. Saeed F, Khokhar A. An overview of multiple sequence alignment. arxiv.org/0901.2747. https://arxiv.org/ftp/arxiv/papers/0901/0901.2747.pdf
3. Wikipedia. Multiple sequence alignments. https://en.wikipedia.org/wiki/Multiple_sequence_alignment

Clustal Omega

4. Sievers F, Wilm A, Dineen D, Gibson TJ, Karplus K, Li W, Lopez R, McWilliam H, Remmert M, Söding J, Thompson JD, Higgins DG. Fast, scalable generation of high-quality protein multiple sequence alignments using Clustal Omega. Mol Syst Biol. 2011;7:539. https://doi.org/10.1038/msb.2011.75.
5. Sievers F, Higgins DG. Clustal omega. Curr Protoc Bioinformatics. 2014;48:3.13.1–3.13.16. https://doi.org/10.1002/0471250953.bi0313s48.

Chapter 9
Identifying Conserved Sequence Motifs

Sequence motifs are integral components of biological sequences, representing conserved patterns that are important for the function and regulation of protein function. Motifs also occur in DNA where they are commonly associated with regulatory elments such as transcriptional promoters and terminators. In this chapter, we will identify the conserved sequence motifs and examine the relationship between the protein's three-dimensional structure and conserved sequence motifs.

9.1 Objectives

- Identify protein functional motifs using a variety of motif identification resources.
- Compare the sequence motifs found with different resources to the multiple sequence alignment constructed in Chap. 8.
- Compare conserved motifs to the three-dimensional structure of the protein.

9.2 Background

The dictionary definition of a motif is "a recurrent thematic element in an artistic or literary work." We use motif similarly when we talk about protein sequences and structures—a motif is a recurrent sequence or structural element.[1] Motifs are common in proteins because of their descent from common ancestors, and because of functions that are common to diverse proteins. For instance, ATP binding, DNA binding, and formation of membrane pores are common types of protein motifs. The other cause of motifs is the combined effects of mutation and selection—changes in

[1] The American Heritage Dictionary of the English Language. Houghton Mifflin Harcourt, 2016.

© The Author(s), under exclusive license to Springer Nature Singapore Pte Ltd. 2025

S. Wang, M. Gribskov, *Practical Bioinformatics*,
https://doi.org/10.1007/978-981-96-7949-2_9

sequence in regions that are critical to function are removed by selection, giving rise to prominently conserved regions in multiple alignments, as we saw in Chap. 8.

Because motifs are directly and causally related to protein structure and function, we are interested in identifying motifs, especially in novel sequences with unknown function. Finding previously known motifs in new sequences lets us make a hypothesis about the function of the protein and is one of the main steps in genome annotation. On the other hand, identifying motifs with no known function is useful because it suggests that a novel activity may have been discovered. One way to identify the function associated with such unknown motifs is to examine the functions of all proteins that contain a similar motif.

Both of these approaches require methods to define motifs, for convenience we'll call them patterns, and to match the motif definitions with sequences. There are three basic approaches, each one more complex but also more powerful than the last: consensus sequences, regular expressions, and position-specific scoring matrices (PSSM). The oldest and weakest method is the consensus sequence method. Consensus sequences were first introduced to define DNA sequence motifs such as promoters and operators. The definition is very simple—a consensus sequence is a string of letters that represents the majority rule character at each column of an MSA. Matching to consensus sequences is also simple—just slide the motif across a sequence and note the positions where the letters of in the consensus sequence match all the letters in the sequence of interest. The problem is that consensus sequences lose important information present in the MSA; in particular, they do not distinguish positions that are completely conserved (inference: positions that are critical for function) and positions where one of the letters has only a slight plurality (inference: not very important for function). The big advantage of consensus sequences is that they are simple to display and remember. Senior molecular biologists often know dozens of consensus sequences. Because of the simplicity and clarity of displaying consensus sequences, they are often shown even when matching sequences are identified by more powerful methods.

Regular expressions (RE) are an improvement over consensus sequences and can represent all consensus sequences (they are a superset of consensus sequences). RE are still a string representation, but they allow more than one letter to match at each position. This retains much more of the information present in the MSA and distinguishes those columns where multiple letters are observed in the MSA. RE also allow positions to be unconserved, that is, to match to any letter in a sequence (typically represented as N in nucleic acid sequences or X in protein sequences). However, some information is still lost; although two letters are present at a particular column in an MSA, one may be much more dominant than the other (indicating its greater functional importance). RE also allow a limited amount of variation in the spacing between conserved letters (which would be seen as insertions in the MSA). The big advantage of RE is that they have been the subject of intense study in computer science for decades, and very efficient methods are known for identifying them in large texts (such as genome databases).

Position-specific scoring matrices (PSSM), also known as position weight matrices or profiles, can represent all consensus sequences and RE, making them the most inclusive pattern description. In fact, they retain all first-order information (information that belongs to single residues rather than pairs or larger groups). PSSM can be thought of as a matrix where the rows correspond to the columns of an MSA, and the columns correspond to all the letters that could be found in the sequence: the 4 nucleotide bases in the case of nucleic acids and the 20 amino acid residues in the case of proteins. The values in the PSSM may be simply the probabilities of seeing each letter at each position, or they may be fit to biological models of sequence letter similarity. Another advantage of PSSM is that they can be matched to sequences allowing gaps, and the match scores follow the same distribution as matches between sequences (an extreme value distribution, Chap. 5). The ability to calculate *P*-values and *E*-values for matching PSSM and sequences greatly increases our ability to identify structurally and functionally important regions of novel proteins (keep in mind than novel protein sequences are being identified as an exponentially increasing rate as more genomes are sequenced).

Many databases of protein motifs have been created, each using different motif definitions, but generally falling into the three classes described above. Most of these definitions have been gathered into a single encyclopedic resource, InterPro, that we will use in this lab. InterPro contains all three kinds of motifs discussed above: consensus sequences, RE, and PSSM, as well as a great deal of information describing the structure and function that can be expected of sequences that match and a comprehensive list of known match proteins.

One well-known protein sequence motif is the zinc finger motif. It is a protein structural motif that allows proteins to interact with DNA, RNA, or other proteins. It is named for its coordination of zinc ions, which stabilize the structure. A zinc finger consists of a short peptide that folds around a zinc ion, forming a compact, finger-like structure. The consensus sequence for a zinc finger motif often includes a specific pattern of cysteine and histidine residues that coordinate the zinc ion. A common type of zinc finger motif is the Cys2His2 type, which has the following consensus sequence: C-X(2,4)-C-X(3)-[LIVMFYWC]-X(8)-H-X(3,5)-H (C represents cysteine, H represents histidine, X represents any amino acid). In this format, which is used by the PROSITE resource, square brackets indicate sets of amino acid residues that can match and numbers in parentheses represent the number of allowed amino acid that can be repeated at a certain position; X[3], for instance indicates the presence of three amino acid residues of any type. Zinc finger motifs are found in a wide variety of proteins, including many transcription factors, where they are used to recognize specific DNA sequences and regulate gene expression. They are also found in proteins involved in RNA processing and protein-protein interactions.

In nucleotide sequences, sequence motifs often indicate the precise locations where proteins, such as transcription factors or RNA-binding proteins, interact with the genetic material to regulate gene expression. For instance, the TATA box is a conserved DNA sequence motif found in the promoter regions of many genes in eukaryotes. It is named for its base sequence, which is typically TATA[AT]A[AT] or a variation thereof, and is located around 25 to 30 base pairs upstream of the transcription start site. The TATA box serves as a recognition site for the TATA-binding protein (TBP), which is a component of transcription factor IID (TFIID). The binding of TBP to the TATA box is a crucial step in the initiation of gene transcription by RNA polymerase II. The TATA box is particularly important in the regulation of genes that are expressed in a tissue-specific or developmentally regulated manner.

Because sequence motifs are directly and causally related to the structure, function, and regulation of macromolecules, motif databases, such as InterPro, are very useful in studying questions such as

1. Gene regulation: Identifying motifs in the regulatory regions of genes can help predict where transcription factors bind and how genes are turned on or off in response to different signals.
2. Protein function and interaction: Sequence motifs can indicate functional domains in proteins that are involved in catalysis, small molecule binding, protein-protein interactions, and binding to DNA or RNA. This information is especially important for identifying the function and molecular mechanisms of proteins identified in newly sequenced genomes.
3. Phylogenetic analysis: Sequence motifs can be used to compare relatedness among different species. Conserved motifs across species may indicate essential functional regions that have been preserved through evolution.

As we discussed in the previous labs, MSAs, in which a number of homologous sequences are simultaneously aligned with each other, play an important role in bioinformatics. One of the important uses of multiple sequence alignments is in constructing phylogenetic trees (see Chap. 10). They also provide a huge amount of information about which regions of a protein family are conserved, and therefore, which regions are most important for its structure and function.

There are a number of ways that protein structure and function give rise to evolutionary constraints. The most general effect has to do with the partitioning of the hydrophilic (water-loving) and hydrophobic (water-fearing) residues between the interior and exterior of the protein structure. The interior of folded proteins is densely packed, comparable to an organic solid, and denser than an organic liquid such as octanol. It is energetically unfavorable to bury polarizable groups in the tightly packed internal core of the protein; that is, a mutation that would bury a polarizable hydrophilic group destabilizes the structure and interferes with protein folding. As an analogy, you can imagine trying to force a polarizable molecule, water, into a hydrophobic solvent such as vegetable oil. We all know that this is difficult, and left to itself, the water and oil will separate. It is even more disruptive to bury charged residues (lysine, arginine, glutamate, and aspartate) in the interior of a protein, except in the special case where a positively charged residue interacts with a negative one to form a salt bridge.[2]

[2]For those who would like a little more detail: The reason for the partitioning of hydrophilic and hydrophobic residues largely has to do with *entropy*, which is a measure of disorder. Hydrophobic molecules in water tend to be surrounded by an organized "cage" of water molecules held together by hydrogen bonds. When a hydrophobic amino acid side chain moves from an aqueous (water) environment to the core of the protein, this cage of water is released, thereby increasing entropy. This positive entropy change has been called the *hydrophobic effect* and is one of the main driving forces of protein folding.

A second constraint arises due to the tight packing of the protein interior—a mutation that substitutes a large residue for a small residue, or vice versa, in the core of the protein tends to destabilize the protein structure. It's easy to see why trying to force a large residue, such as phenylalanine, into the space occupied by a small residue, such as alanine, is difficult. The opposite case, a mutation that replaces a large residue with a small one, is equally disruptive because, without alterations to the folded structure, this would leave a cavity, a vacuum, in the middle of the protein. As we know, this is an unnatural and physically unlikely situation (*horror vacui*).

What we see in a multiple alignment tends to follow these principles; the residues that are on the protein surface tend to be hydrophilic and accumulate more mutational changes. The residues that are buried in the core of the protein tend to be hydrophobic and to accumulate mutations more slowly. We also see that, due to the periodic nature of the alpha helix (Fig. 9.1), a common secondary structural element in proteins, there is often a repetitive pattern of about 3–4 residues in conservation (each turn of an alpha helix requires 3.6 residues). Other elements of protein structure produce their own constraints. For instance, regions in which the amino acid chain reaches the surface and quickly turns back to the interior (a beta turn) typically have three to four hydrophilic residues in a row and include proline residues at the turn. Coiled coil proteins, in which two long alpha helices wrap around each other, often contain a characteristic repeat of H-P-P-H-C-P-C (H = hydrophobic, P=polar, and C = charged), the so-called heptad repeat. In

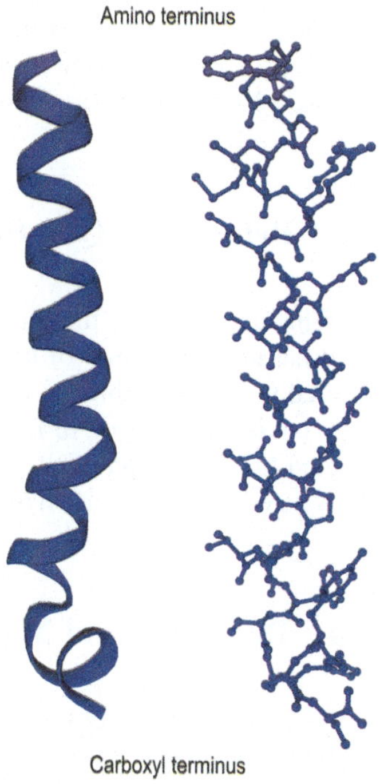

Fig. 9.1 Protein alpha helix. The alpha helix is a repetitive secondary structure held together by hydrogen bonds between the carbonyl oxygen of each residue i and the amide nitrogen of residue $i + 4$

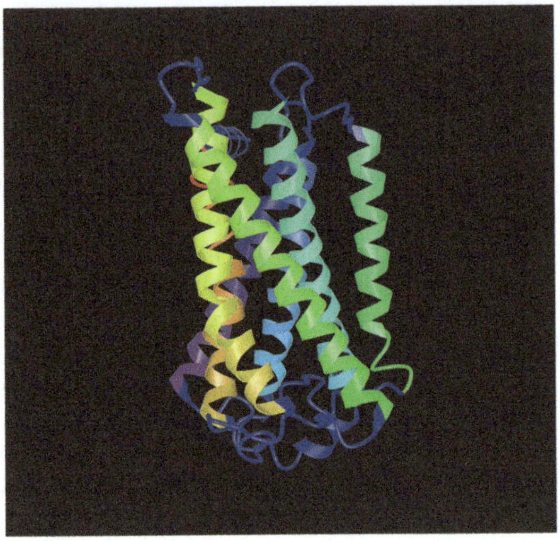

Fig. 9.2 The transmembrane region of follicle stimulating hormone receptor. The colored regions are transmembrane, the solid blue are not

some cases, constraints arise due to interactions between protein molecules and other cellular molecules, such as nucleic acids and lipids (Fig. 9.2).

In this chapter we will investigate the relationship between the conserved sequence regions and the structure of the protein. As you recall, these conserved regions arise because mutations in structurally important regions are more frequently removed by the selection process than are mutations in unimportant regions. We will use the NCBI iCn3D structure viewer to compare the sequence and structure of the Mcr-1 protein. This viewer has the advantage of running directly from a web browser. You will perform a similar analysis using one of your aoi proteins.

9.3 Before Lab Questions

- What are sequence motifs? Why are they important?
- What is the meaning of hydrophilic, and which amino acid residues are considered to be hydrophilic?
- What is the meaning of hydrophobic, and which amino acid residues are considered to be hydrophobic?
- Where in a protein are hydrophobic amino acid residues found?

9.4 Resources and Materials

- NCBI/Blast website: https://blast.ncbi.nlm.nih.gov/Blast.cgi
- InterPro website: https://ebi.ac.uk/interpro/

- Eggnog website: http://eggnog6.embl.de
- Sequence of Colistin resistance protein Mcr-1 (supplemental file 3).
- iCn3D structure viewer—https://www.ncbi.nlm.nih.gov/Structure/icn3d/full.html?mmdbid=5K4P
- Protein sequence of your aoi.

9.5 Activity 1: Identify Sequence Motifs Using CDD

Use the Blast protein search you performed earlier (Chap. 6) for this section. You may be able to find your previous result by clicking the *Recent Results* tab, but if not, simply repeat the search. If repeating the search, please change the database to *Protein Data Bank proteins*; this will make the search faster because this database is smaller. We will use the MCR-1 protein as an example, but you will need to use one of your aoi protein sequences to look for motifs. It is simplest to use the same protein family that you used in Chap. 8.

Step 1: In your BlastP result, select the *Graphic* Summary tab. The conserved domain diagram at the top of the result (Fig. 9.3) shows the location of known protein domains based on the CDD database. In this diagram, the black bar indicates the query sequence, and the colored bars below indicate known protein domains or functional domains. Clicking on the bars in the diagram brings up an expanded diagram (Fig. 9.4). Each of the bars in this diagram are clickable and leads to more detailed information. Below the diagram, there is a table of E-values indicating the strength of the match to each motif. Remember, small E-values are better, so these matches to MCR-1 are all quite strong. The CDD results are divided into high confidence hits, labelled *specific hits,* and hits that are statistically significant but lower scoring which are labeled as *nonspecific hits*.

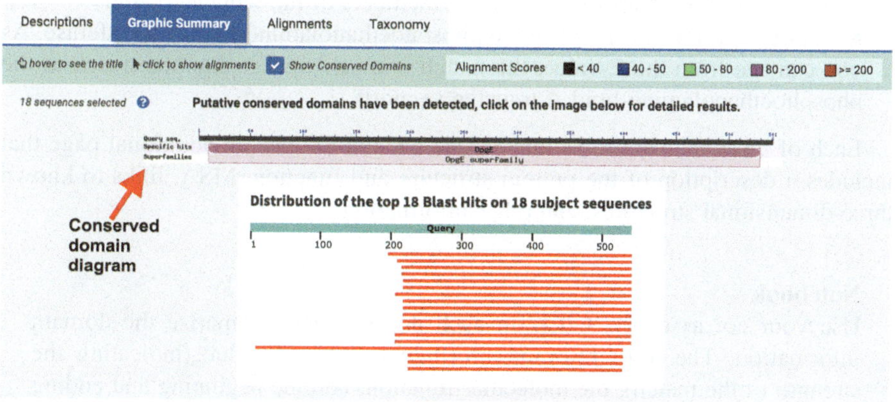

Fig. 9.3 Blast result graphic summary tab showing the CDD diagram at the top

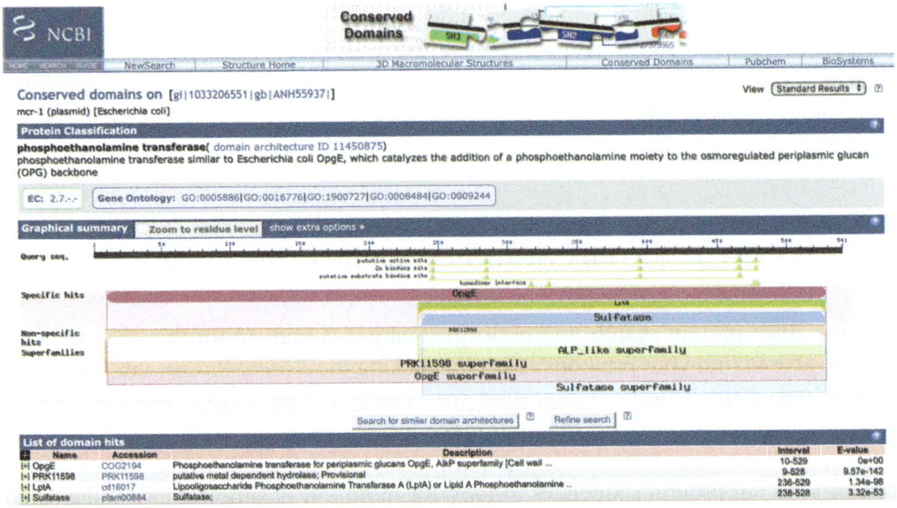

Fig. 9.4 Conserved domains (https://www.ncbi.nlm.nih.gov/Structure/cdd/cdd.shtml) search results for colistin resistance protein mcr-1

For colistin, we see the following specific hits:

- P_eth_NH2_trans—phosphoethanolamine-lipid A transferase. This is a sequence superfamily implicated in colistin resistance.
- OpgE—phosphoethanolamine transferase for periplasmic glucans (OPG), alkaline phosphatase superfamily, another protein sequence family related to phosphoethanolamine-lipid A transferase. Note that the length of this domain is much shorter and the E-value of the match is higher (worse).
- LptA—lipooligosaccharide phosphoethanolamine transferase A (LptA) or lipid A phosphoethanolamine transferase. This is essentially the same as the first superfamily but has more information about members of the sequence family.
- Sulfatase—Sulfatases are related to phosphoethanolamine-lipid A transferase. As you can see from the E-value, the match is much weaker than the match to the phosphoethanolamine-lipid A transferase motif.

Each of these bars are links that can be clicked to open an additional page that includes a description of the protein structure and function, MSA, links to known three-dimensional structures, and other information.

Notebook

Use your aoi as query. Click on each domain and summarize the domain information. The most relevant information is the E-value (indicating the strength of the match), the molecular function, and the beginning and ending

(continued)

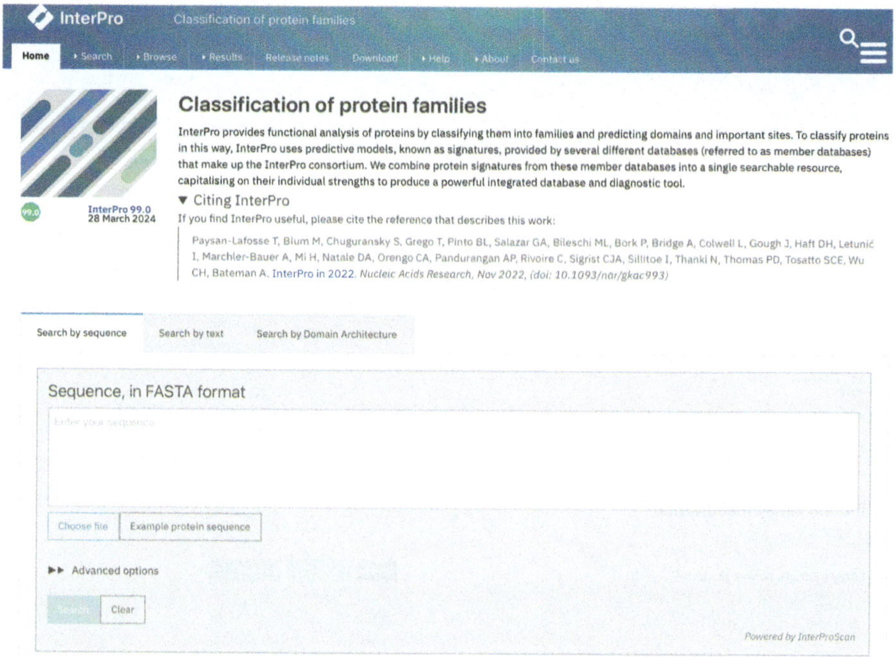

Fig. 9.5 InterProScan submission page (https://www.ebi.ac.uk/interpro/)

positions of each domain in the query sequence. Is your aoi related to more than one group of enzymes based on the CDD results? If so, are different regions of the aoi protein related to different conserved domains, or are the multiple hits in the same region? What would each of these possibilities indicate?

9.6 Activity 2: Identify Motifs Using InterPro

Step 1: Open the InterProScan[3] submission page (Fig. 9.5), and copy your aoi protein sequence into the query box (labeled *Enter your sequence*).

Step 2: Submit the search and wait for a few seconds for the result. Click on the title of the search to open the result page, which should look something like (Fig. 9.6). The result is divided into sections:

[3]O'Cathail C, Ahamed A, Burgin J, et al. The European Nucleotide Archive in 2024. Nucleic Acids Research. 2025 Jan;53(D1):D49-D55. DOI: https://doi.org/10.1093/nar/gkae975.

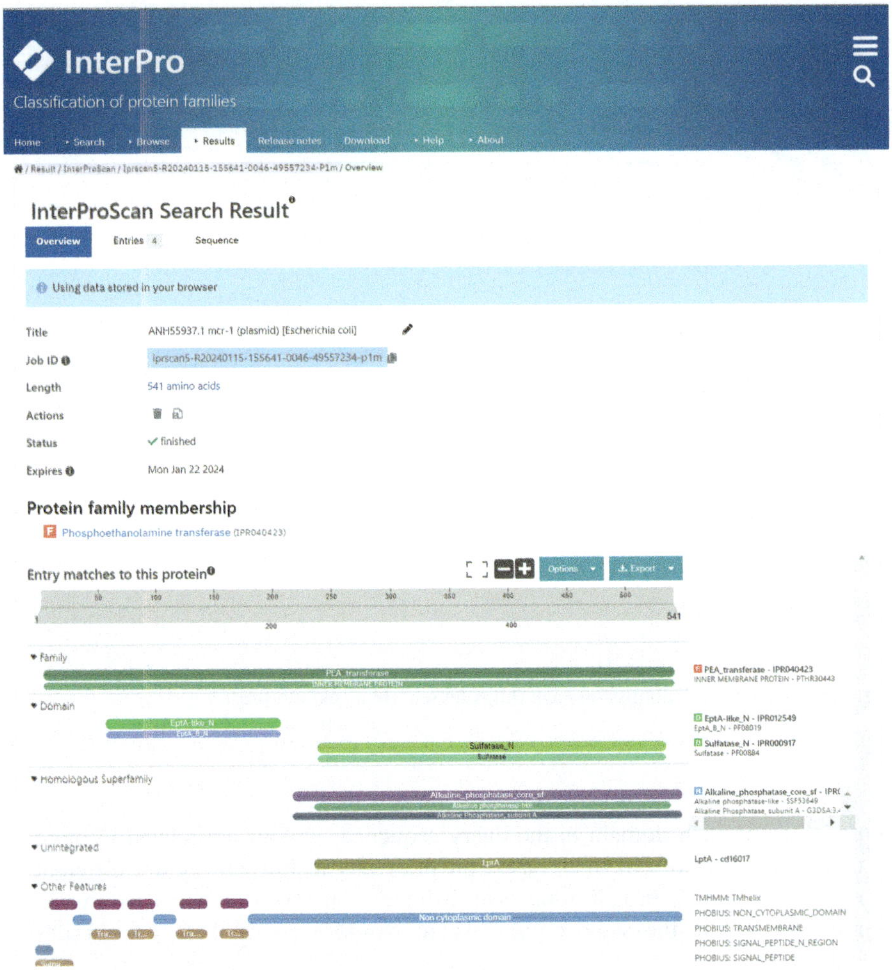

Fig. 9.6 InterProScan Results page

- Family—Matches to gene families.
- Domain—Matches to known protein domains. Domains are typically autonomous regions of genes/proteins that are capable of functioning on their own.
- Homologous superfamilies—Gene/protein families that share the same domain structure.
- Unintegrated—As InterPro grows, new motifs/domains are added and their annotations are integrated with known features. Unintegrated simply indicates that the features are present but not yet integrated.

- Other features—The main information in this area is predicted transmembrane regions, with the corresponding predications of the external and cytoplasmic regions of the protein, and predicted signal peptides (which target proteins for export from this cell). These predictions are highly reliable and provide useful information about the location of the protein in the cell.
- Residues/other residues—indicates individual amino acid residues involved in the active site, ligand binding, or posttranslational modification.
- InterPro/Panther GO terms—Gene ontology terms that can be associated with this sequence based on the matches to motifs in the database. The quality of these annotations is variable, so consider these as suggestions rather than high-quality predictions.

Step 3: Interpreting results. Immediately under the title, *InterProScan Search* Result, are three links; by default, the *Overview* is shown, but the *Entries* result is often simpler and easier to understand. In the *Entries* view, click on the Interpro Accession ID (for instance, IPR004423) or the feature name (*for instance, PEA_transferase*) to see more information. In the *Overview,* each panel represents the query sequence with the position of matching features/motifs/domains shown as colored boxes. Hovering over a box will show additional information. At the right of the diagram, the title of the feature/motif/domain is shown, and clicking on this will show you the complete entry for this feature in the InterPro database. The database entry gives a brief description of the function of the feature and provides literature references. It also indicates other features that overlap or are commonly found in association. The left-hand column of the database entry gives links to known proteomes, protein structures (experimental and AlphaFold predicted), pathways, and taxonomic groups in which the feature is found. A tabular form of the results can be obtained by clicking on the *Export* tab. This output has the exact coordinates of the matching features in the query, as well as the E-value for each match that provides one. This is very useful for evaluating the significance of the match. I find the GFF or TSV formats to be easiest to interpret.

Notebook

Record significant matches to protein families, features such as transmembrane domains, and modified, binding, and active site residues. Does this entry give any hints about the origin of the Mcr-1 colistin resistance gene? Can you suggest a possible origin for your aoi protein?

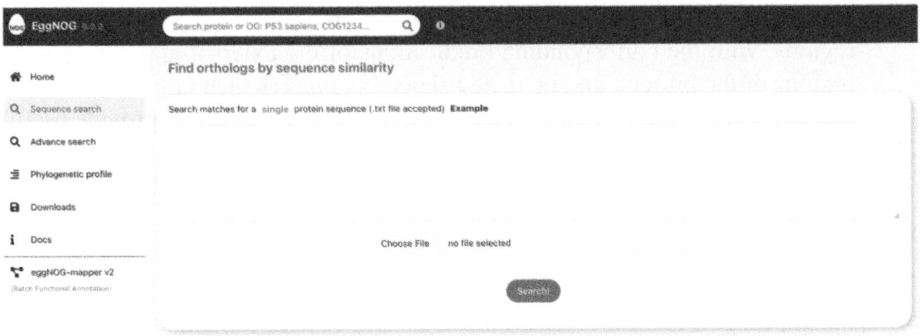

Fig. 9.7 EggNOG sequence search submission page (http://eggnog6.embl.de)

9.7 Activity 3: Identify Colistin Resistance-Related Proteins Using EggNOG

Eggnog is a database of proteins computationally predicted to be orthologs. Eggnog v6.0 (April 2024) contains 17 million orthologous protein groups derived from 12,535 organisms including prokaryotes, eukaryotes, and viruses. EggNOG[4] can be searched by entering the sequence/gene name on the main page or by sequence by selecting the *Sequence search* option in the left menu and cutting and pasting the MCR-1 sequence in the search box. Searching with a sequence is usually more reliable.

Step 1: Open the EggNOG submission page (Fig. 9.7), click sequence search in the left panel, copy the MCR-1 protein sequence into the query box, and click *search.*

Step 2: As in most database searches, you will get many matches. The match with the smallest E-value is generally the most relevant, especially if the E-value is less than 1e-50 or so. The results contain a wealth of information, including the names of matching ortholog families, matching functional domains, and gene symbols. This information significantly overlaps with the information in CDD and InterPro. EggNOG provides unique information about the taxonomic groups in which the orthologous group is found: a MSA, and a premade phylogenetic tree.

In the result for colistin (Fig. 9.8), we see results with E-values of 6.4e-164, 8.9e-118, and 1.3e-28. The entry with the lowest E-value is labelled *lipid A ethanolamine phosphotransferase*, which we know from our previous search is the name of the Mcr-1 resistance protein. This entry is also found in the most species and has the most protein examples—both are clues that this is the most interesting entry.

[4] Ana Hernández-Plaza, Damian Szklarczyk, Jorge Botas, Carlos P Cantalapiedra, Joaquín Giner-Lamia, Daniel R Mende, Rebecca Kirsch, Thomas Rattei, Ivica Letunic, Lars J Jensen, Peer Bork, Christian von Mering, Jaime Huerta-Cepas, eggNOG 6.0: enabling comparative genomics across 12,535 organisms, Nucleic Acids Research, Volume 51, Issue D1, 6 January 2023, Pages D389–D394.

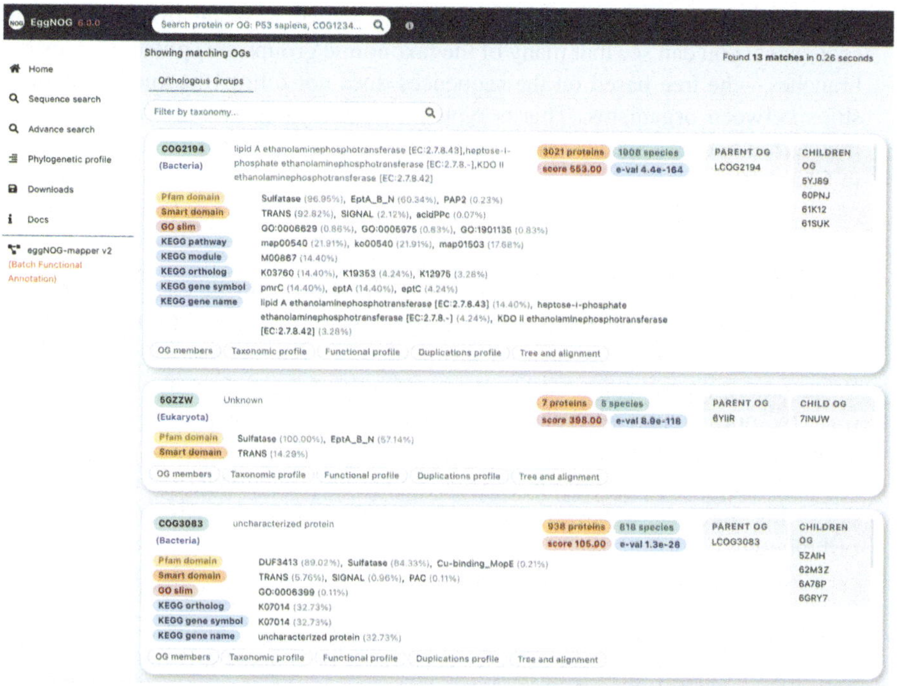

Fig. 9.8 EggNog search results

> **Notebook**
> For each matching orthologous family, make a table in your notebook showing how many orthologs of your aoi protein are found in each taxonomic phylum (not superphylum). Record the summary information describing the orthologous families in your notebook.

Step 3: Examine the taxonomic profile. You can see the number of orthologs in each taxonomic group by clicking on the ring. In the case of colistin, we see that the protein is widespread in proteobacteria, but is also found in the FCB group (a superphylum including the *Fibrobacterota, Chlorobiota,* and *Bacteroidota phyla), Terrabacteria*, and to a much lower extent in other groups.

Step 4: Return to the main result page and select *Tree and alignment*. The branch lengths in the phylogenetic tree shown on the left are proportional to the sequence distance. On the right, the regions of the sequence matching functional domains are shown (in a sense, this is an alignment, but not what is normally meant). For colistin, we can see that with only a few exceptions, every protein contains two domains, an EPT_B_N domain and a sulfatase domain. These correspond to the domains found previously but have been named differently. More information on these domains can be found by returning to the main entry and looking in the list of domains. If you look closely at the tree, which is made from the protein sequences, you can see that many of the taxonomic groups are split among distant branches—the tree based on the sequences does not follow the known relationships between organisms. This is typical of genes that have been horizontally transferred between species. We will look more closely at horizontal gene transfer in Chap. 10.

Notebook
Include the phylogenetic tree/functional domain diagram in your notebook. Note that you can control the appearance of the tree using the *Control panel* button. You will probably have to shrink the tree considerably to make it fit. What pattern do you expect for a gene/protein that is being inherited vertically (*i.e.*, by normal parent-to-child inheritance)? Is this the pattern you see? Include specific examples.

Do all major branches of the tree match the same functional domains? What does this indicate about the function of the protein across the whole orthologous group?

Step 5: Return to the main result page and select *Functional profile*. The functional profile tabulates the number of times each functional domain is found across the entire functional group. Domains that occur in a majority of orthologs are the ones that most strongly associated with the group.

Notebook
Record the functional domains most associated with this orthologous group and the percentage of group members they are found in.

Compare and discuss the results obtained with InterPro, CDD, and EggNOG. Make a table summarizing the residue positions and likely functions associated with the different parts of the sequence.

Are they consistent with each other? Which features/motifs/domains are unique to only one of the resources used.

9.8 Activity 4: Comparing Motifs with the Three-Dimensional Structure

The structure and function of proteins places a strong evolutionary constraint on the evolution of proteins. These constraints are the origin of the conserved sequence features that we looked at in activities 1–3. In this activity we examine the relationship between the protein's three-dimensional structure and conserved sequence motifs. Since the most conserved regions detected in the multiple sequence alignment correspond to the most functionally important regions of the protein, we are interested to know if they correspond to any of the sequence motifs. We will walk you through the process of comparing the conserved regions in the MSA with the crystallographic structure of the protein. Several crystallographic structures are available for Mcr-1. The one we will look at is named 5K4P by the protein database (PDB) and contains the coordinates of a homodimer, which is thought to be the active form. For technical reasons relating to protein crystallization, only the structure of the catalytic region of the molecule, residues 218–541, is available; The N-terminal region, which interacts with the membrane, was removed from the protein before structure determination. You have been provided with the PDB file, which gives the three-dimensional coordinates of every atom in the structure, a total of 2931 atoms. In addition to the protein, the structure contains water molecules, zinc atoms, and sorbitol. The sorbitol molecule is in roughly the position you would expect to see the substrate lipopolysaccharide bound. The zinc atoms are known to be important for the function of this enzyme. Follow the example, and do the same with the structure of your aoi protein. In the case of MCR-1, the crystalized protein contains only residues 218–541 of the original MCR-1, so we limit our focus to just this region. For your aoi, you will compare either an actual three-dimensional structure or a predicted structure to the conserved regions you see in the multiple sequence alignment.

The sequences in the multiple sequence alignment have been extended by the addition of gaps, so the residue positions in your MSA and sequence logo do not correspond directly to the MCR-1 sequence. The only way I know to get the coordinates in terms of the MCR-1 sequence is to find the corresponding regions in the MSA, and from that, the position in the original sequence (using the FastA sequence file).

Step 1: First you need to find a 3D structure that corresponds to your aoi. There are several ways to do this. First, look in UniProt under the structure category (in the left-hand menu bar). Under the 3-D structure databases heading you may find an experimentally determined (crystallographic) structure. Usually these will be listed as links to PDB, PDBe, RCSB-PDB. If a PDB identifier is listed you can use in directly. However it is useful to download the three-dimensional coordinate file as well by clicking on one of the links. Generally, structures that are more complete (cover a wider range of amino acid residues) or have better resolution (smaller number in Å) are preferred. Alternatively, you may see a structure listed with the source "AlphaFold". This is a computationally predicted structure, which is often quite accurate, and can be used in place of an experimentally predicted structure. A third way to find a structure is to do a blast search vs the PDB database at NCBI, using your original aoi protein sequence as a query. Using the structure viewer is easiest if you can identify a PDB identifier (such as 5K4P for Mcr-1) but you can also use downloaded coordinates (for structures predicted by AlphaFold). If you use the URL https://www.ncbi.nlm.nih.gov/Structure/icn3d/full.html?mmdbid=5K4P, it will open with the MCR-1 structure (5K4P.pdb) displayed. If you are using a PDB coordinate file or can't see a view of the structure, use the *File* menu to load the PDB coordinate file. Initially, you will

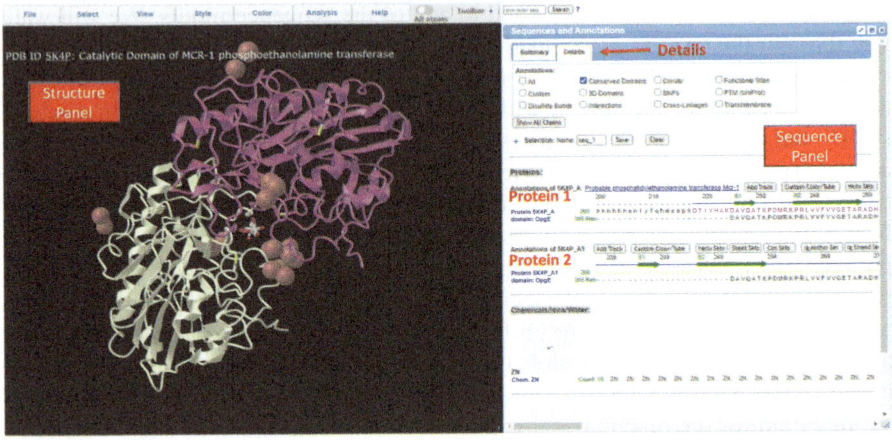

Fig. 9.9 iCn3d view of the 5K4P(MCR-1) structure and sequence. (https://www.ncbi.nlm.nih.gov/Structure/icn3d/full.html?mmdbid=5K4P)

see the protein chains as a ribbon diagram of their backbone, in different colors, the zinc atoms as spheres, and sorbitol as a stick figure.

Click on *Analysis*, then click *Seq & annotations* to show the protein sequence in a panel to the right (Fig. 9.9). The initial view will show you an overall summary with the protein molecule(s) shown as a ribbon and non-protein molecules shown as a stick model, or as balls. In this case, the bound sorbitol molecule is shown as a stick, and the bound Zinc ions as pink balls. Selecting the *Details* tab under *Sequences and Annotation* will show you the actual amino acid residue sequence and the residues that interact with the small molecules (sorbitol and Zn) in the structure. This particular structure includes two copies of the Mcr-1 protein (shown here in beige and pink). The bar at the bottom of the panel is used to move the sequence to the right and left.

In the left panel, the sequence can be rotated by holding the left mouse button and moving the mouse. Move the mouse very slowly or the image will spin out of control. Hold the right mouse button to translate the image, and use the scroll wheel to zoom in and out. With practice you should be able to focus on any part of the structure.

ICn3D provides many options for depicting the protein and small molecule structures. These range from renderings of the molecular surface, to space filling atomic models, to more abstract depictions such as the initial ribbon view. These options can be selected using the Style tab. Styles are applied to the currently selected part or parts of the structure, which can be selected using the Select tab, or by left clicking and moving the mouse across the sequence in the right panel. Note that molecular surfaces are computationally intensive to calculate and rotation will be very jerky (suggestion: don't calculate surfaces until you have your final view set up). Other visualizations, such as showing active site residues as ball and stick models don't effect rotation. Several built-in selections are particularly useful: *Main Chains* will select just the protein molecules (not the small molecule ligands); *by Distance* lets you select atoms around another selected regions such as active sites; *Select on3D* lets you select regions in the display using the mouse. The latter options let you see the interface between, for instance, a small molecule such as sorbitol and the protein molecule. After you have selected a region it will be highlighted (the default color is yellow but you can change it using the *Color* tab). In the sequence panel you can see both the sequence itself (both molecules in the structure), and the secondary structure on the row above it (S indicates beta strand, H alpha helix, and c indicates other irregular structure). Below the sequence, individual lines show the protein residues that interact with other molecules. In this structure, these are the 10 Zn atoms, two sorbitol molecules, and the other protein chain. ICn3D provides many ways to highlight information about the structure. Some of these are in the Color menu. Spectrum colors the protein chain with a spectrum of color from N-terminus to C-terminus. *Secondary* colors the chain according to secondary structure. *Hydrophobicity* colors the chain according to the hydrophobicity of the amino acid side-chains.

Try some of these options and practice rotating the molecule (hold left mouse button and move mouse; translating the molecule, right button and move mouse; and zooming with the mouse wheel; if you have a touch pad use the equivalent movements). Initially only the protein backbone is shown as a ribbon diagram.

Step 2: The sorbitol molecule in this structure is located approximately at the substrate binding site. Let's see what parts of the sequence are close to the sorbitol ligand.

(a) Check that *Select on 3D* is set to *Residue* in the Select menu.
(b) Rotate and zoom until you can clearly see the sorbitol molecule.
(c) For PC users, select the sorbitol using *Alt-left-click.* For Mac users, select the sorbitol using *option-left-click.* You should now see an outline around the sorbitol (the default color is yellow).
(d) In the Select menu, choose by Distance. For "1. Select the first set" choose "selected". For "2. Sphere with a radius" enter "6 Å". Finally, for "3. Select the second set" choose "non-selected" and click Display. The selected region outlined in yellow should now show all the atoms within 6 Å of the sorbitol ligand. You can't see much detail because only the protein backbone is displayed (Fig. 9.10).
(e) Examine the sequence window and see what sequence residues are selected. These are the amino acid residues close to the sorbitol pseudo-substrate. You should see two widely separated regions, one around residue 85 and another around residue 285.

Step 3: Now, let's see what parts of the sequence are close to the Zinc atoms. There are many Zinc atoms, so we'll look at just one cluster.

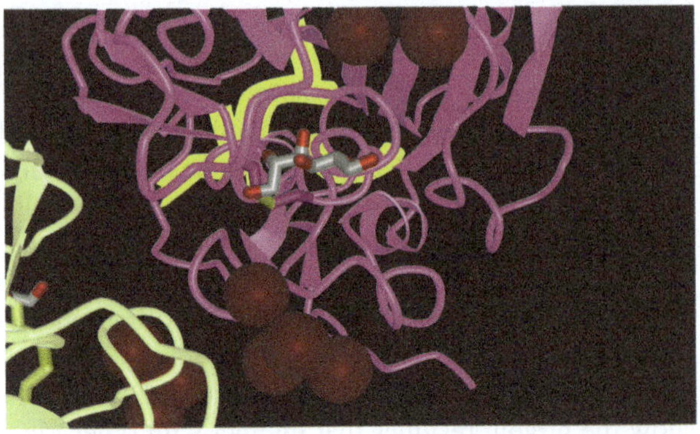

Fig. 9.10 Residues around bound sorbitol selected in structure

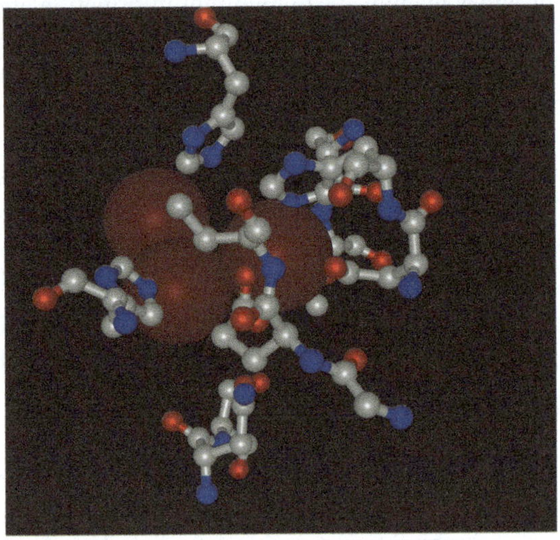

Fig. 9.11 Screenshot from mcr-1 structure (https://www.ncbi.nlm.nih.gov/Structure/icn3d/full.html?mmdbid=5K4P). Zn10-Zn12 Cluster showing several interacting histidine residues. Zinc atoms are shown as red spheres

(a) In the sequence window, select Zn10, Zn11, and Zn12 by dragging over the Zn labels while holding down the mouse. These atoms will be shown with an outline in the left hand structure panel.

(b) Select a sphere with a radius of 6 Å of the zinc cluster using the same process as in step 1d, above. Even though the *Distance* appears to be selected, you have to click it again. The same for the *Display* button. You can see the amino acid side chains that interact with the zinc cluster by selecting *Style -> Proteins -> Stick* (Fig. 9.11.). We also change the color, select *color - > spectrum.* Zn has a tendency to interact with histidine and cysteine residues. Do you see any in this region?

Step 4: For your aoi, in the sequence window, select each of the conserved regions you found in the MSA (Chap. 8). Examine where these regions lie in the structure. Do any of them seem to be related to either the substrate (which would be close to the active site) or in the case of MCR-1 to the Zn atoms (which we know are involved in catalysis)? You should have some idea of relevant active site residues or other important features from your initial research on your aoi protein. If nothing is known about the catalytic mechanism of your aoi, are the conserved regions grouped in the three dimensional structure (this might be the location of the protein active site).

9.9 Analysis/Notebook

- Record the amino acid residues within 6 Å of the active site of your aoi and whether these correspond to any of the conserved sequence regions. Convert the numbering of the amino acids to agree with the complete aoi protein sequence, not the numbering shown by iCn3D. Do the same for any other important molecules in the structure, such as the sorbitol of Zn atoms in the MCR-1 structure.
- For the conserved regions recorded in your table of conserved motifs, describe where these regions fall in the structure and what their functional significance may be.

Further Reading

1. Proteins: secondary structure and fibrous proteins. http://cbc.arizona.edu/classes/bioc462/462a/NOTES/Protein_Structure/secondary_structure.htm

Chapter 10
Building a Phylogenetic Tree

In this chapter, we will use sequences of your aoi gene family from multiple species of bacteria to build a phylogenetic tree, and compare the tree made based on these sequences to the known taxonomy of bacteria. In a real study, we would use the nucleotide sequence of the genes, but for simplicity, we will use the protein sequences in this lab. Protein sequences are easier to unambiguously align, and we don't have to worry about red herrings such as differences in GC content. This makes the construction of the tree simpler. We will use a simple distance-based method for tree construction. When the sequences from very distant species are closer in the tree than they should be, it is good evidence for horizontal gene transfer.

10.1 Objectives

- Understand basic distance-based methods for tree construction.
- Understand the difference between gene trees and species trees.
- Apply tree construction to your aoi.
- Use the aoi tree to decide if horizontal gene transfer is an important feature of resistance to your aoi.

10.2 Background

Phenetic and Cladistic Trees

One of the long-term goals of biology has been to produce the tree of life—a diagram that shows how every organism is related to each other. Beginning with Linnaeus, trees were constructed based on the morphological characteristics of organisms: presence of hair, feathers, or scales; homeothermy vs poikilothermy, tail vs no-tail, and more complicated quantitative measures, such as the ratio of the length of the

© The Author(s), under exclusive license to Springer Nature Singapore Pte Ltd. 2025
S. Wang, M. Gribskov, *Practical Bioinformatics*,
https://doi.org/10.1007/978-981-96-7949-2_10

skull to the width. Such trees, in which organisms are classified based on overall similarity, are called *phenetic* trees. At the end of the twentieth century, there was a huge debate between advocates of *phenetics* and advocates of *cladistics*. A cladistic classification attempts to group organisms according to their ancestral relationships, with more closely related organisms in narrower systematic groups. One of the primary principles of a cladistics classification is *monophyly*, the idea that all members of a taxonomic grouping should be closer to each other than they are to any outside groups. Such a group, which comprises all organisms descended from a common ancestor, is also called a *clade*. This seems like an obvious and familiar principle; all primates are more closely related to each other than any are to canids or equids. Clearly, the traditional Linnaean classification system, with its division into kingdoms, phyla, classes, orders, etc., contains many polyphyletic groups and is not cladistic. For instance, we know that mammals are descended from a reptile, and reptiles from fish, but the traditional classification makes fish, reptiles, and mammals all classes of equal rank.

The problem with using morphology (overall physical similarity) as a basis for classification is that it is difficult to weigh the importance of one feature vs another. Which is more important, for instance, the presence of five toes or the presence of fur? Thus, when classifications are based on morphological characteristics, it is not that easy to distinguish the phenetic and cladistic perspectives. The debate between phenetics and cladistics, came to a head when the advent of sequencing methods made it possible to correctly determine ancestral relationships based on DNA and protein sequences. One of the reasons sequences are so much more powerful is that, to a first approximation, all sequence characters are on the same scale—there is much less need to weight the relative importance of characters (*i.e.*, nucleotide bases and amino acid residues). Another advantage of sequence information is that hundreds or even thousands of characters (the DNA bases or amino acid residues) are available.

Constructing Distance Trees

Conceptually, constructing a tree based on a matrix of distances is simple. Initially, every sequence (taxon) is considered to be in its own group. The general approach alternates between finding the two taxa with the shortest distance (greatest similarity) and joining them into a single group. Followed by recalculating the distances between the new group and the other taxa. These steps are then repeated until all the taxa have been joined into a single group (sometimes called an Operational Taxonomic Unit or OTU).

A more detailed description of the construction of distance-based trees requires several steps:

Step 1: Full-length sequences spanning the phylogenetic range of interest must be selected. This will be done using Blast searches.

Step 2: An MSA will be constructed based on the protein sequences. This alignment allows us to calculate the distance between each pair of sequences. The distance can be simply the number of positions at which the aligned sequences differ.

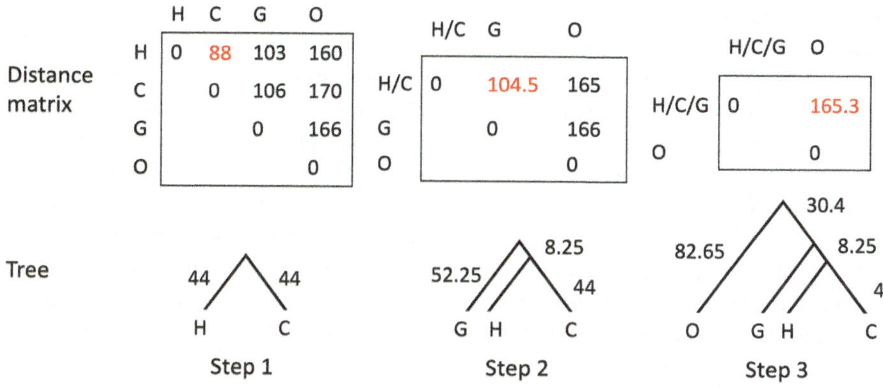

Fig. 10.1 Construction of a distance-based tree. The top row shows the original distance matrix (based on nucleotide differences between aligned hemoglobin sequences) and the updated distance matrix as each pair of OTUs is joined. The shortest distance is highlighted in red. The bottom row shows the corresponding tree as each OTU is added. Note that the distances on the branches are half of the distance between the OTUs

Distances can also be calculated from the similarity score for the alignment based on an empirical scoring tables such as BLOSUM62 (see Sect. 6.2). Because mutations can occur more that once at the same site, distances are frequently adjusted for such multiple substitutions.

Step 3: Initially, every sequence (taxon) is considered to be in its own group. Then the two taxa with the shortest distance (greatest similarity) are found and joined into a single group or OTU, (Fig 10.1).

Step 4: The distances between the new OTU and the other taxa are then recalculated. Most simply this is done by simply averaging the distances from the two OTUs being joined to each of the other OTUs.

Steps 3 and 4 are then repeated until all the taxa have been joined into a single OTU. Fig. 10.1 shows an example of this approach applied to Human (H), Chimpanzee (C), Gorilla (G), and Orangutan (O) hemoglobin sequences. In this case, we use a method called unweighted pair group method of averages (UPGMA), which recalculates the distances at each step as the simple average of the distance to each taxon being joined into the new OTU. Note that in real life, the distance between the root (at the top of the tree) and the leaves (H, C, G, O) will not usually be equal as they are in this example (Fig. 10.1) because of different evolutionary rates along the branches of the tree.

There are many other possible ways to calculate/update distances: One of the most useful called is Neighbor-Joining (NJ). Without going into the details of the calculation, it suffices to say that UPGMA does not take full advantage of the information in the distance matrix – for instance, in calculating the distance of the branch from the H/C ancestor to H and C, the information contained in the distances between G and H and G and C is not used. NJ uses this additional information and therefore makes better trees.

Horizontal Gene Transfer

In this laboratory, we will investigate the importance of horizontal transfer in the spread of antibiotic resistance, once again using Colistin as an example. Horizontal transfer of antibiotic resistance genes on plasmids between different species of bacteria is one of the most problematic aspects of the spread of antibiotic resistance. Many resistance plasmids carry resistance to multiple antibiotics, and the acquisition of such a plasmid instantly creates a multiple drug-resistant bacterium. As discussed in Chap. 2, such plasmids can be transferred between species, and even phyla. In the presence of widespread overuse of antibiotics, such multidrug-resistant plasmids become common in both pathogenic and non-pathogenic bacteria in the general environment, creating reservoirs from which they can easily reenter pathogenic bacteria.

Our approach in this laboratory is to use sequences of your aoi resistance family (Chaps. 4 and 6, 7, 8) from multiple species of bacteria and compare the tree made based on these sequences to the known taxonomy of bacteria (Fig. 10.2).[1] The taxonomy keeps updating, so this figure might not be the up-to-date by the time this book is published.

When the MCR-1 sequences are phylogenetically closer than their host bacterial species are known to be, it is good evidence for horizontal transfer. In a real study, we would use the nucleotide sequence of the genes, but for simplicity, we will use the protein sequences. Protein sequences are easier to unambiguously align, and we don't have to worry about red herrings such as differences in GC content. This makes the construction of the tree simpler. We will use a distance-based method for tree construction. Although distance methods are not the method of choice for making professional phylogenetic trees,[2] they have several key advantages: they are computationally fast, they do not require a high level of expertise, and the number of parameters that need to be investigated and tuned is very small (essentially zero).

Because of the high interest in bacterial antibiotic resistance, resistance genes have been sequenced in many different bacteria, often in thousands of different species. This allows us to look at whether these genes are spreading through normal (vertical) inheritance or through lateral (horizontal) inheritance. Many antibiotic resistance genes have been annotated in the databases, so all we have to do is look them up. The approaches were illustrated in previous chapters. To be specific, we need to:

---

[1]Mendler K, Chen H, Parks DH, Lobb B, Hug LA, Doxey AC. AnnoTree: visualization and exploration of a functionally annotated microbial tree of life. Nucleic Acids Res. 2019 May 21;47(9):4442–4448.

[2]Distance methods are not wrong, but they have largely been superseded by maximum likelihood methods, which incorporate many more parameters into the evolutionary model (for instance, variable evolutionary rates, and differences between rates on different branches of the tree). While these newer methods are powerful, using them requires considerable theoretical and computational expertise. In spite of their less sophisticated approach, thousands of papers using distance methods appear every year.

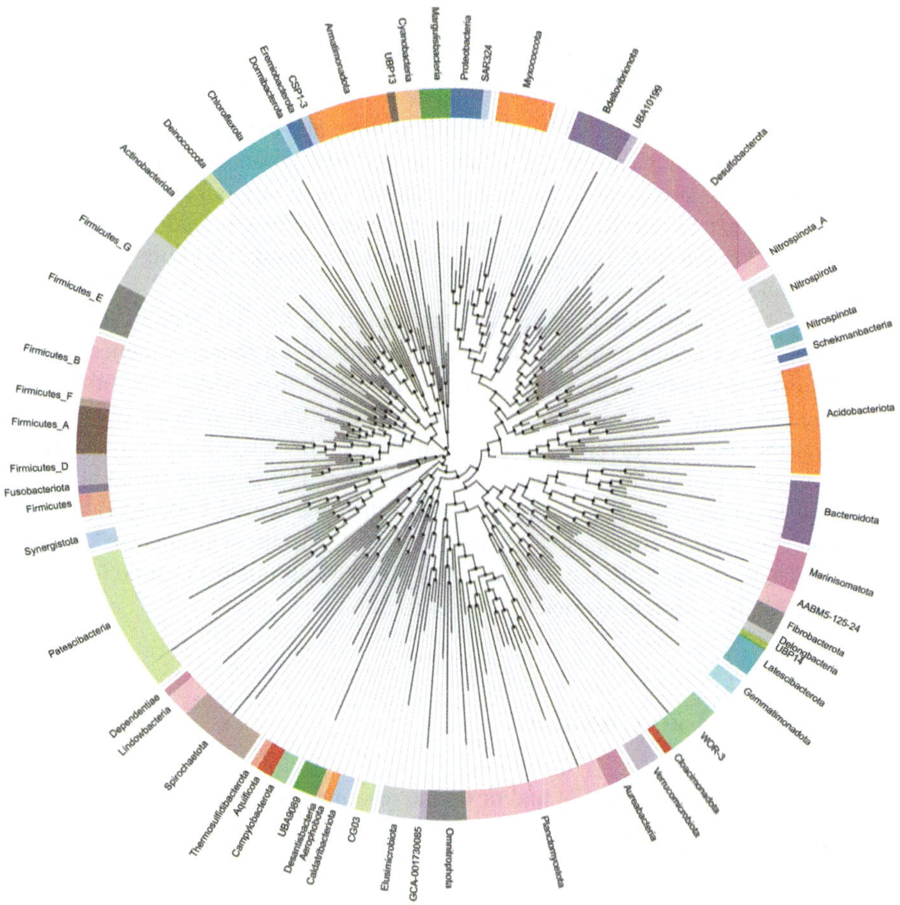

Fig. 10.2 Known taxonomy of bacteria. The tree was made by AnnoTree (http://annotree.uwaterloo.ca/annotree/app/). The parameter settings producing this figure: resolution level is "class," tree type is "bacterial," taxonomic level is "phylum")

1. Identify a set of widespread antibiotic resistance genes (specific for a single resistance mechanism).
2. Construct a phylogenetic tree that shows the relationship between the sequences.
3. Compare the sequence-based tree to the known bacterial taxonomy.

The result should show one of two possible patterns.

Pattern 1: The sequence-based tree is mostly the same as the known taxonomy. In this case the gene is being transferred vertically—the sequences in closely related species are more similar because those species more recently diverged from a common ancestor.

Pattern 2: The sequence-based tree is discordant with the known taxonomy, with sequences from different families or even phyla appearing to be very similar. This is strong evidence for horizontal transmission of the genes.

Over a hundred thousand species of bacteria have had their genomes sequenced. This information has been compiled into an updated taxonomy for bacteria and archaea, which is available online in the genome taxonomy database (http://gtdb.ecogenomic.org/). Figure 10.2 shows the known bacterial tree, including genera with more than 60 known species. This will be your gold standard in looking for evidence of horizontal transfer.

In this lab, our approach is to use sequences from the MCR-1 gene from multiple species of bacteria and compare the tree made based on these sequences to the known taxonomy of bacteria. We will also construct a MSA, which is similar to the tree made in Chap. 8. However, the goal here is different. Instead of trying to maximize the sequence diversity of the sequences, in this lab, we want to maximize the taxonomic diversity in order to see whether the MCR-1 proteins found in widely separated taxonomic groups are more similar than we would expect. Table 10.1 shows an abbreviated taxonomy for bacteria. Nearly all of these groups are known to have the MCR-1 protein. Since there are only about 55 families, it is quite feasible to identify a MCR-1 protein sequence from each family.

10.3 Before Lab Questions

Using the introductory material and/or online resources, define the following:

- Morphological character
- Phenetic tree
- Monophyletic
- Clade
- Horizontal transfer

Answer the following questions:

1. What is the objective of a cladistic phylogenetic analysis?
2. Why is the traditional Linnaean classification not considered cladistic?

10.4 Resources and Materials

- Relabeled protein sequences for aoi from Chap. 8.
- Clustal Omega—https://www.ebi.ac.uk/jdispatcher/msa/clustalo
- Simple Phylogeny—https://www.ebi.ac.uk/jdispatcher/phylogeny/simple_phylogeny
- TreeDyn—https://www.phylogeny.fr/one_task.cgi?task_type=treedyn

Table 10.1 Known bacterial taxonomy based on GTDB

Phylum	Class	Order	Family	Genus
Actinobacteriota	*Actinobacteria*	*Actinomycetales*	*Actinomycetaceae*	*Actinomyces*
			Bifidobacteriaceae	*Bifidobacterium*
			Microbacteriaceae	*Microbacterium*
			Micrococcaceae	*Rothia*
		Corynebacteriales	*Corynebacteriaceae*	*Corynebacterium*
				Gordonia
				Mycobacterium
				Nocardia
				Rhodococcus
			Micromonosporaceae	*Micromonospora*
		Propionibacteriales	*Nocardioidaceae*	*Nocardioides*
			Propionibacteriaceae	*Cutibacterium*
		Streptomycetales	*Streptomycetaceae*	*Streptomyces*
Bacteroidota	*Bacteroidia*	*Bacteroidales*	*Bacteroidaceae*	*Bacteroides*
				Prevotella
			Porphyromonadaceae	*Porphyromonas*
			Tannerellaceae	*Parabacteroides*
			UBA932	*RC9*
		Flavobacteriales	*Flavobacteriaceae*	*Flavobacterium*
			Weeksellaceae	*Chryseobacterium*
				Elizabethkingia
		Sphingobacteriales	*Sphingobacteriaceae*	*Sphingobacterium*
Campylobacterota	*Campylobacteria*	*Campylobacterales*	*Arcobacteraceae*	*Arcobacter*
			Campylobacteraceae	*Campylobacter*
			Helicobacteraceae	*Helicobacter*
Cyanobacteriota	*Cyanobacteriia*	*Synechococcales_A*	*Cyanobiaceae*	*Prochlorococcus*

(continued)

Table 10.1 (continued)

Phylum	Class	Order	Family	Genus
Firmicutes	*Bacilli*	*Bacillales*	*Anoxybacillaceae*	*Geobacillus*
			Bacillaceae	*Bacillus*
		Bacillales_A	*Planococcaceae*	*Lysinibacillus*
		Lactobacillales	*Enterococcaceae*	*Enterococcus*
			Lactobacillaceae	*Lactobacillus*
				Leuconostoc
				Oenococcus
			Listeriaceae	*Listeria*
			Streptococcaceae	*Lactococcus*
				Streptococcus
		Mycoplasmatales	*Ureaplasmataceae*	*Mycoplasma*
		Staphylococcales	*Staphylococcaceae*	*Staphylococcus*
	Bacilli_A	*Paenibacillales*	*Paenibacillaceae*	*Paenibacillus*
Firmicutes_A	*Clostridia*	*Clostridiales*	*Clostridiaceae*	*Clostridium*
		Lachnospirales	*Lachnospiraceae*	*Ruminococcus*
		Peptostreptococcales	*Peptostreptococcaceae*	*Clostridioides*
Fusobacteriota	*Fusobacteriia*	*Fusobacteriales*	*Fusobacteriaceae*	*Fusobacterium*
Patescibacteria	*Paceibacteria*	*UBA9983*	*Nomurabacteraceae*	*UBA9973*
Proteobacteria	*Alphaproteobacteria*	*Acetobacterales*	*Acetobacteraceae*	*Acetobacter*
		Caulobacterales	*Caulobacteraceae*	*Brevundimonas*
		Rhizobiales	*Beijerinckiaceae*	*Methylobacterium*
			Rhizobiaceae	*Agrobacterium*
				Bartonella
				Brucella
				Mesorhizobium
				Rhizobium

				Sinorhizobium
			Xanthobacteraceae	*Bradyrhizobium*
		Rhodobacterales	*Rhodobacteraceae*	*Paracoccus*
		Rickettsiales	*Rickettsiaceae*	*Rickettsia*
		Sphingomonadales	*Sphingomonadaceae*	*Sphingobium*
				Sphingomonas
				Sphingopyxis
	Gammaproteobacteria	*Betaproteobacteriales*	*Burkholderiaceae*	*Achromobacter*
				Bordetella
	Gammaproteobacteria	*Betaproteobacteriales*	*Burkholderiaceae*	*Burkholderia*
				Cupriavidus
				Ralstonia
			Neisseriaceae	*Neisseria*
		Ectothiorhodospirales	*Thioalkalivibrionaceae*	*Thioalkalivibrio*
		Enterobacterales	*Aeromonadaceae*	*Aeromonas*
			Alteromonadaceae	*Alteromonas*
				Idiomarina
				Pseudoalteromonas
			Enterobacteriaceae	*Citrobacter*
				Cronobacter
				Enterobacter
				Erwinia
				Escherichia
				Gilliamella
				Klebsiella
				Leclercia
				Pantoea
				Pectobacterium

(continued)

Table 10.1 (continued)

Phylum	Class	Order	Family	Genus
				Proteus
				Salmonella
				Serratia
				Yersinia
			Pasteurellaceae	*Haemophilus*
				Mannheimia
				Pasteurella
			Shewanellaceae	*Shewanella*
			Vibrionaceae	*Photobacterium*
				Vibrio
		Francisellales	*Francisellaceae*	*Francisella*
		Legionellales	*Legionellaceae*	*Legionella*
		Pseudomonadales	*Hahellaceae*	*Marinobacter*
			Halomonadaceae	*Halomonas*
			Moraxellaceae	*Acinetobacter*
				Moraxella
				Psychrobacter
			Pseudomonadaceae	*Pseudomonas*
		Xanthomonadales	*Xanthomonadaceae*	*Stenotrophomonas*
				Xanthomonas
Spirochaetota	*Leptospirae*	*Leptospirales*	*Leptospiraceae*	*Leptospira*
	Spirochaetia	*Borreliales*	*Borreliaceae*	*Borreliella*
		Treponematales	*Treponemataceae*	*Treponema*
Verrucomicrobiota	*Chlamydiia*	*Chlamydiales*	*Chlamydiaceae*	*Chlamydia*
				Chlamydophila

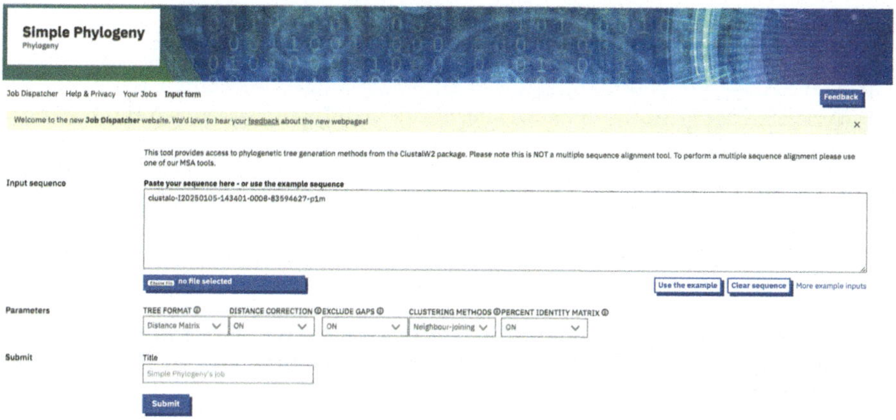

Fig. 10.3 Simple Phylogeny submission page (https://www.ebi.ac.uk/jdispatcher/phylogeny/simple_phylogeny)

10.5 Activity 1: Constructing the Phylogenetic Tree

We will use the Simple Phylogeny program[3] (Fig. 10.3) to make a tree.

Step 1: Upload the multiple sequence alignment results obtained from Clustal Omega.

Step 2: Modify the settings. Make the following changes in the "STEP 2—Set your Phylogeny options" section (Fig. 10.3).

- TREE FORMAT—select *Distance Matrix* to get a report of the calculated distance between each pair of sequences. The distance matrix is provided as an additional output file under the *Result Summary* tab.
- DISTANCE CORRECTION—*on*. Adjusts calculated distances for multiple mutations at the same site.
- EXCLUDE GAPS—*on*. Gaps in the multiple sequence alignment represent insertion/deletion events in one or more sequences. Alignments in the region of gaps are sometimes less reliable and contribute noise to the tree.
- CLUSTERING METHOD—*Neighbor-joining*. You may try UPGMA if you like. The method described in the introduction of the lab is the UPGMA method. If you use UPGMA, include both NJ and UPGMA trees in your notebook.
- P.I.M—*on*. The percent identity between every pair of sequences, as a matrix, is provided as an additional output file under the *Result Summary* tab.

[3]Madeira F, Madhusoodanan N, Lee J, et al. The EMBL-EBI Job Dispatcher sequence analysis tools framework in 2024. Nucleic Acids Research. 2024 July;52(W1):W521–W525.

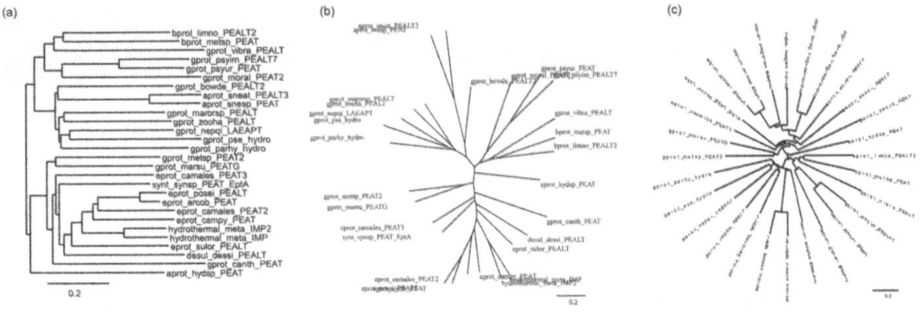

Fig. 10.4 Phylogenetic Trees Visualized by TreeDyn (https://www.phylogeny.fr/one_task.cgi?task_type=treedyn). (**a**) Rectangular. (**b**) Radial. (**c**) Circular. These trees are drawn based on exactly the same Newick input

The first part of the output shows the tree, corrected for multiple mutations and excluding gaps, is shown in a serial text format known as Newick format.[4] The program shows a visualization of the tree but, at least in my case, was squashed into a very narrow space; it was hard to make out the details. To get an easier-to-analyze tree, select and copy the *Newick tree* and proceed to the next activity.

10.6 Activity 2: Visualizing the Phylogenetic Tree

To examine our tree, we will visualize it using an online program called *TreeDyn*.[5] Paste the Newick-formatted version of your tree into the box and press select. You should see a nicely formatted version of your tree. You can cut and paste this by selecting the image (right-click on PC), or select a variety of formats to download in the bar below the image. TreeDyn allows you to rearrange the tree without changing the way the branches are connected and can also show radial and circular trees (which some people prefer). Examples of different tree visualizations are shown in Fig. 10.4. Each of these trees shows exactly the same relationships - study the trees to understand the differences in the representation.

This tree is still difficult to analyze because it does not contain taxonomic information about the sequences. If you edited your original sequence file to have phylogenetically informative labels you should be able to see them in the tree. If you did not, the easiest approach is to cut and paste your tree into PowerPoint and add the taxonomy labels by hand (see example in Fig. 10.5).

[4]Consult the Wikipedia article on *Newick format* if you'd like to know the details of how it works.

[5]Dereeper A, Guignon V, Blanc G, Audic S, Buffet S, Chevenet F, Dufayard JF, Guindon S, Lefort V, Lescot M, Claverie JM, Gascuel O. Phylogeny.fr: robust phylogenetic analysis for the non-specialist. Nucleic Acids Res. 2008 July 1;36(Web Server issue):W465–W469.

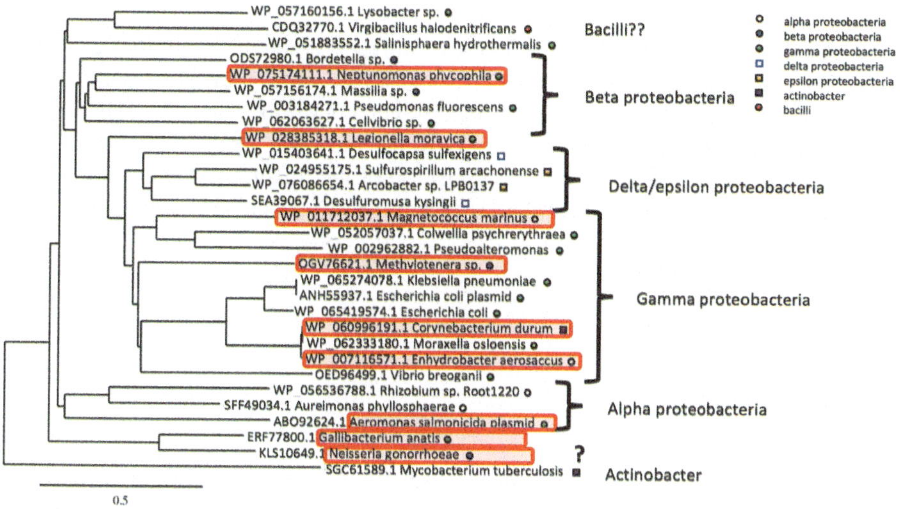

Fig. 10.5 Sample annotated tree. Taxa marked in red are misplaced in the tree. Taxonomic groups were assigned by inspection. For instance, *Bacilli* and *Acinetobacter* lie outside the *Proteobacteria* and therefore should branch off first. This is true *for Acinetobacter*, but the placement of the *Bacilli* is unclear because I included only one sequence. Within the proteobacteria, we expect *the Alphaproteobacteria* to branch off first, and the Beta, Gamma, and Delta/Epsilon *Proteobacteria* to each form a branch. Branches were assigned by looking at where the majority of sequences within the class are located

You can use TreeDyn to color the labels in your phylogenetic tree by their respective phylum or class. To do this, select *Colorize*, found in the section *Select an action and click leaf or internal branch* located below the tree plot, choose a color, and add a label, for example, *Alphaproteobacteria*. Then, select the sequences in your tree that belong to that group by clicking on the label (leaf) in your tree. When you are ready to color-code sequences from a different phylogenetic group—change the color and the label, and continue "coloring" your tree. Note that you can color a branch and all of its children in one operation by checking *branch* and unchecking *leaf.*

Notebook

- Record the colored multiple sequence alignment in your Notebook. The alignment is the basic data for the tree. To repeat your results, another investigator would need this alignment.
- Record the Newick-formatted tree (raw numerical data) and the initial cladogram provided by Clustal Omega. The Newick tree is the basic data for the tree. If anyone wanted to repeat your results, they would need this information.

(continued)

- Carefully examine your tree to see whether it agrees with the known bacterial taxonomy. Discrepancies suggest horizontal transfer of your chosen antibiotic resistance gene, probably via a plasmid. Mark these discrepancies in your tree. You do not have to use the same approach as I used in the example, but try to make the position of the misplaced taxa clear.
- From your tree, can you explain why antibiotics resistance become a severe problem?

Further Reading

1. http://evolution.berkeley.edu/evolibrary/article/evo_03 – An online introduction to the basics of evolutionary analysis and phylogenetic trees.

Chapter 11
Gene Expression Analysis Using RNA-Seq Data

Gene expression can be defined as the transcription from DNA to messenger RNA followed by translation into proteins. With the development of next-generation sequencing technology (NGS), it has become possible to quickly determine the complete sequence of both the genome and transcriptome of an organism. RNA sequencing (RNA-Seq) can discover novel transcript variants and splice sites, or quantify mRNAs for gene expression analysis. Implicitly, since mRNA transcripts encode the entire set of proteins in an organism, we can also obtain the organism's proteome. RNA-seq has become a powerful tool to study gene expression in biological labs. However, analysis of such high-throughput sequencing data requires certain programming skills and statistical knowledge; we will introduce the basic steps in RNA-seq analysis, the multiple testing problem, and how to analyze the differentially expressed genes in R software in this chapter.

11.1 Objectives

- To become familiar with the workflow of RNA-seq.
- To understand the issue of multiple testing in RNA-seq.
- To apply DESeq2 package to identify differentially expressed genes.

11.2 Background

Messenger RNA is translated into proteins and the levels of many proteins are determined by the amount of mRNA that is present. In higher eukaryotic organisms, a gene can have different alternatively spliced forms or isoforms when it has multiple exons, which increases the complexity of the mRNA population and contributes to protein diversity. Note that not all genes are transcriptionally regulated; genes may

© The Author(s), under exclusive license to Springer Nature Singapore Pte Ltd. 2025
S. Wang, M. Gribskov, *Practical Bioinformatics*,
https://doi.org/10.1007/978-981-96-7949-2_11

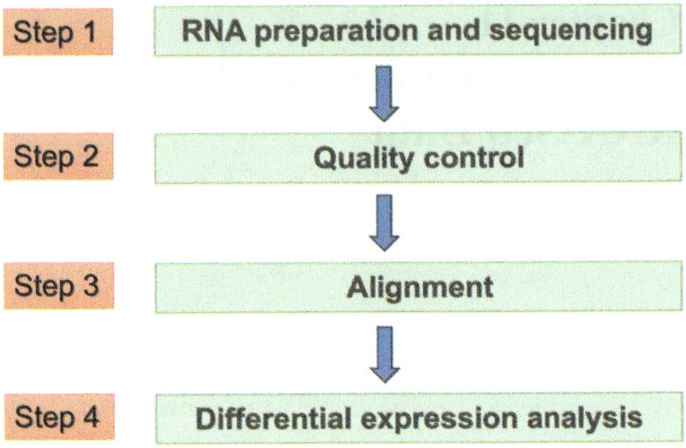

Fig. 11.1 The workflow of RNA-seq

also be regulated at the level of translation, or posttranslationally by protein modification, or by interaction with other molecules.

RNA-Seq is a next-generation sequencing (NGS) technology that directly determines the mRNA sequence by first converting the RNA to a double stranded cDNA sequence. Compared with quantitative real-time PCR, RNA-seq has several advantages, for example, the level of all transcribed RNAs in the cell can be simultaneously determined, allowing the detection of both known and novel genes in a single experiment.

There are four steps in an RNA-seq experiment (summarized in Fig. 11.1). The first step is sample preparation and sequencing. The mRNA is selected and then reverse-transcribed to produce a cDNA copy. This is followed by fragmentation to generate consistent-length fragments, to which sequencing adapters are attached, creating a cDNA library. In the most common approach, short-read sequencing (such as Illumina) is used to obtain hundreds of millions of sequencing reads, usually around 150 bases long. The second step is quality control of the raw sequence reads, in which low-quality sequences and sequences containing adapters from the library are removed. Other contaminants, such as host DNA (for a pathogen) or pathogens, may also be removed.

To quantify the number of transcripts corresponding to each gene, the sequence reads must be aligned to a reference genome or transcriptome. The reference genome can be obtained from a completely sequenced and annotated genome. From the reference genome annotation, the set of all known and predicted RNAs transcribed from the genome can be obtained. A reference transcriptome can also be obtained by ab initio assembly of the RNA-Seq reads into transcripts (also known as transcriptome assembly). Aligning reads to the reference genome/transcriptome determines the positions where each read matches the reference. The number of reads that map to a particular transcript is proportional to the amount of RNA in the cell (which depends on both the number of copies of the mRNA per cell and the length of the gene) and can defines the expression level for each gene or isoform. The quality control and alignment steps are often done on a high-performance computer cluster because of the large amount of data produced by NGS.

After the expression level of each gene or transcript is determined, we use statistical analysis to identify differentially expressed genes (DEG) or isoforms (step 4). Many statistical methods have been proposed; some of the most commonly used include edgeR,[1] DESeq2,[2] Limma[3], etc. RNA-Seq experiments are very sensitive to the experimental design. One must think about the experimental design first. What is the treatment? How many replicates do we need, and what are the variables (tissue, condition, or time)? In an RNA-seq experiment, we assume most genes do not change; in other words, the null hypothesis is that the mean difference of gene expression between the control group and the treatment is zero. Clearly this condition cannot be me in many conditions, for instance when cells are infected by a pathogen and are dying, so more complicated analyses than the ones described here must be used.

In a single RNA-seq experiment, we apply a statistical test for differential expression to thousands of genes simultaneously. This creates a multiple testing problem. The threshold α in a statistical test measures the probability of seeing a positive result (differential expression) when there are only random differences in the treatment and control conditions. α is thus a *P*-value, which is defined as the probability of seeing a difference greater than a threshold when no significant difference is present (*i.e.*, the datasets being compared have only random differences). If only random variation is present, all of the genes we predict to be differentially expressed are false positives. Consider an organism with 20,000 transcripts, if we choose $\alpha = 0.05$, we expect to predict that 1000 genes are differentially expressed, even if the true number is zero. What if the true number of differentially expressed genes is 100? In that case, the fraction of genes, out of the ones we predict as positives, that are correct is 100 out of 1100 or 91%. The fraction of genes incorrectly predicted is 91%. This fraction is called the false discovery rate (FDR). In most cases, it's clearly unacceptable for 91% of our results to be false positives.

To reduce the FDR, we adjust the calculated *P*-values for multiple testing, so that our results achieve a specific FDR. FDR thresholds between 0.25 and 0.05 are common. Increasing the FDR always increases the number of false positives as well as the number of true positives detected (called "recall" in the statistical literature). Lowering the FDR decreases the number of false positives, but more true positives are also missed. There is no a priori best FDR threshold; an appropriate threshold depends on the cost of missing a true positive *vs* the cost of having a false positive. There are several methods that can control the false discovery rate (FDR), such as the Benjamini-Hochberg method and the Bonferroni method. Generally, the application of FDR control methods to data will be reported as FDR, adjusted *P*-value, or Q-value.

In differential expression experiments, we measure two distinct aspects of the difference between the sample groups we are comparing (for instance a treatment group and an untreated control group). The first is a statistical effect, represented by the adjusted *P*-value, which estimates the probability that we can distinguish the expression level of a gene in our two groups. The other, log fold-change (LFC), is

---

[1] Robinson MD, McCarthy DJ, Smyth GK. edgeR: a Bioconductor package for differential expression analysis of digital gene expression data. Bioinformatics. 2010 Jan 1;26(1):139–40. https://doi.org/10.1093/bioinformatics/btp616.

[2] Love MI, Huber W, Anders S. Moderated estimation of fold change and dispersion for RNA-seq data with DESeq2. Genome Biol. 2014;15(12):550. https://doi.org/10.1186/s13059-014-0550-8.

[3] Ritchie ME, Phipson B, Wu D, Hu Y, Law CW, Shi W, Smyth GK. limma powers differential expression analyses for RNA-sequencing and microarray studies. Nucleic Acids Res. 2015 Apr 20;43(7):e47. https://doi.org/10.1093/nar/gkv007.

the logarithm of the ratio of the expression level of the same gene in the two groups. Generally, the genes that show the largest (positive) and smallest (negative) LFC are the ones most directly related to the difference in treatments for our two groups. For this reason, LFC is often referred to as the "biological effect."

Once we have identified the significant DEG, we use a "volcano plot" to show the relationship between the "statistical effect" and the "biological effect" (Fig. 11.2). The X-axis in the volcano plot is the log-transformed fold change, the Y-axis is the negative log of the adjusted *P*-value. Each dot represents a gene; genes with negative $\log_2$ fold change show decreased expression compared to the control group. These genes downregulated or turned off in the experimental group, Genes with positive $\log_2$ fold change are upregulated in the treatment compared to the control group. Thresholds for LFC are generally chosen so that most of the genes are not predicted to be differentially expressed (remember, this was a premise of the analysis). In Fig. 11.2, the threshold was chosen to be $\log_2 >= |2|$ (more than 4-fold change up or down).

In this lab, we will analyze RNA-seq data from a mouse cell line experiment, which we provide as a count matrix (the number of reads mapped to each gene). This data was obtained from Illumina reads mapped to the mouse reference genome. In this example the control and treatment groups each comprise three replicates so there are six samples in all. We will learn how to read data using R and how to analyze DEG using the DESeq2 package[4].

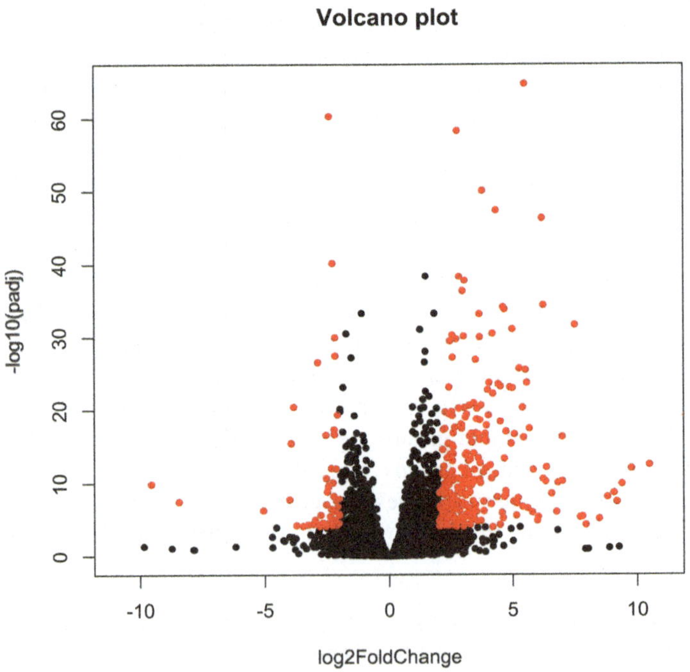

Fig. 11.2 An example of volcano plot. Genes that are differentially expressed are labeled in red color

[4]Love MI, Huber W, Anders S. Moderated estimation of fold change and dispersion for RNA-seq data with DESeq2. Genome Biol. 2014;15(12):550. https://doi.org/10.1186/s13059-014-0550-8.

11.3 Before Lab Questions

Using the introductory material and/or online resources, define the following:

1. Differentially expressed genes (DEG)
2. FDR
3. Volcano plot
4. Log_2 fold change (LFC)
5. Adjusted P-value

11.4 Resources and Materials

- R software (for downloading and installing R, https://www.r-project.org).
- DESeq2 packages (this is the link for installing the package https://www.bioconductor.org/packages/release/bioc/html/DESeq2.html).
- A count matrix: raw.genes.matrix (supplemental file 4).

11.5 Activity 1: Read Data

Open R console and type below commands

```
# Read dataset
# Pay attention to the slash direction in the file path. The default slash in windows computer is toward right, you need to change as shownbelow. "PathToData" means the directory where you save "raw.genes.matrix". Please make sure you put the actual path in below command to read the data.
# read the data in an object called data1
# The count data will be read in and stored in the matrix mycountData.
data1<- read.delim("/PathToData/raw.genes.matrix", stringsAsFactors=F, row.names=1)
head(data1)
dim(data1)
mycountData <- round(data1, digits=0)
head(mycountData)
```

11.6 Activity 2: Use DESeq2 to Identify DEG with Single-Factor Design

```
# DESeq2
# http://www.bioconductor.org/packages/release/bioc/html/DESeq2.html
# Please follow the instructions from the link above to install the DESeq2 package
# after installation, load the package
library(DESeq2)
# give the conditions for each column, in our example data, the first three are control groups,
 then three experimental groups
mycolData <- data.frame(row.names = colnames( mycountData ), condition =
 c( "control","control","control","exp","exp","exp" ))
# generate the data object with the experimental design
mydds <- DESeqDataSetFromMatrix(countData = mycountData, colData = mycolData,
 design = ~condition)
# pre-filtering, filter the genes that have too few reads
sumcontrol<- rowSums(mycountData[,1:3])
sumexp <- rowSums(mycountData[,4:6])
plot(log10(sumcontrol),log10(sumexp))
# from the plot, you can see there are some genes that have the reads only in one condition,
 since we used log10 as the base, in our case, the pre-filtering threshold is set as 100
mydds <- mydds[ rowSums(counts(mydds)) > 100, ]
mydds$condition <- factor(mydds$condition, levels=c("control","exp"))
mydds <- DESeq(mydds)
myres <- results(mydds)
head(myres)
```

```
# Determine threshold, which means how many differentially expressed genes (DEG) will be
kept. We use a combination of adjusted P-value and Log2FC as the filtering conditions.
Sometimes, the number of DEG can be thousands, sometimes only 10 or 20, it depends on the
experimental design and sequencing results. There is no standard or fixed threshold for DEG,
therefore, we need to try different thresholds to see which one gives the reasonable number of
DEG.
resSig <- subset(myres, (padj <= 0.05 & abs(log2FoldChange) >= 1))
dim(resSig)
# it shows this
# [1] 472   6
resdata <- merge(as.data.frame(resSig), as.data.frame(counts(mydds,
 normalized=FALSE)),by="row.names",sort=FALSE)
head(resdata)
# save results in a file called DEG.csv
write.table(resdata,file="/PathToData/DEG.csv",sep=",")
```

11.7 Activity 3: Plot a Volcano Plot

```
plot(myres$log2FoldChange, -log10(myres$padj))
# x axis: log -fold change (biological significance)
# y axis: -log10(adjusted p-value) (statistical significance)
```

11.8 Analysis/Notebook

- In the determine threshold step, use different padj and log2FoldChange mthresholds and compare the results. Make a table of the number of positive and negative DEG you see with different thresholds? What should the threshold be? Explain your reasoning
- The volcano plot is simple; please modify the settings such as adding axis label or changing font size, you may use "?plot" in R for help or try "ggplot2" package, and produce a nicer volcano plot.

Further Reading

1. Wang Z, Gerstein M, Snyder M. RNA-seq: a revolutionary tool for transcriptomics. Nat Rev. Genet. 2009;10(1):57–63. https://doi.org/10.1038/nrg2484.
2. Stark R, Grzelak M, Hadfield J. RNA sequencing: the teenage years. Nat Rev. Genet. 2019;20 (11):631–56. https://doi.org/10.1038/s41576-019-0150-2.
3. McDermaid A, Monier B, Zhao J, Liu B, Ma Q. Interpretation of differential gene expression results of RNA-seq data: review and integration. Brief Bioinform. 2019;20(6):2044–54. https://doi.org/10.1093/bib/bby067.
4. Love MI, Huber W, Anders S. Moderated estimation of fold change and dispersion for RNA-seq data with DESeq2. Genome Biol. 2014;15(12):550. https://doi.org/10.1186/s13059-014-0550-8.
5. Intro to R and RStudio for Genomics. https://datacarpentry.org/genomics-r-intro/
6. False discovery rate: simple definition, adjusting for FDR. https://www.statisticshowto.com/false-discovery-rate/
7. The Benjamini-Hochberg procedure (FDR) and P-value adjusted explained. https://www.r-bloggers.com/2023/07/the-benjamini-hochberg-procedure-fdr-and-p-value-adjusted-explained/

Chapter 12
Machine Learning Methods in RNA-Seq

It is often difficult to immediately understand why specific genes are differentially expressed. Once we have identified DEGs, we are interested in identifying groups of genes that respond similarly, or which biological networks are affected. We use machine learning methods to do further analysis.

Machine learning is a research area where computers simulate or learn human behaviors to acquire knowledge or skills, reorganize information, and structure. It is an interdisciplinary field that involves computer sciences, statistics, probability theory, and others. With the continuous increase in demand for data analysis in the era of big data, how to efficiently acquire knowledge through machine learning has become one of the main tasks in research and industry.

There are two machine learning approaches: supervised learning and unsupervised learning. Supervised learning methods produce models based on known classified data (also called labeled data), for instance a group of genes known to be involved in a certain process. Unsupervised methods are applied when we have no gold standard data (*i.e.*, unlabeled data) on which to train a model. Basic differential expression analysis is an unsupervised procedure – we group genes together based on the difference in their expression in two conditions (control and treatment). More complicated experiments, such as those with multiple treatment groups or timecourse experiments are more difficult to interpret. The hope is that we can deduce biological meaning behind the observed differences in a differential expression experiment by grouping similar genes together and looking at them jointly.

In Chap. 11, we identified DEGs. Now we are interested in identifying groups of genes that respond similarly, or which biological networks are affected. In this chapter, we will focus on unsupervised machine learning methods, including cluster analysis and dimensionality reduction (principal component analysis, PCA). We will learn how to cluster genes and visualize them as heatmaps and how to do PCA in R.

© The Author(s), under exclusive license to Springer Nature Singapore Pte Ltd. 2025

S. Wang, M. Gribskov, *Practical Bioinformatics*,
https://doi.org/10.1007/978-981-96-7949-2_12

12.1 Objectives

- Know the difference between supervised learning and unsupervised learning.
- Understand cluster analysis and principal component analysis.
- Learn how to perform cluster and principal component analysis in R.

12.2 Background

In the era of big data, machine learning has become a very important and critical supporting technology for data analysis and data mining. With the explosion of data, the volume of data has increased exponentially, and new types of data are constantly emerging, such as text analysis, image recognition, and analysis of biological network data. With the rapid production of genomic and other types of sequencing data, the focus of bioinformatics research is gradually shifting from accumulating data to interpreting these data. New discoveries in biology will greatly rely on the analytical ability to combine and correlate diverse data across multiple dimensions and scales. For example, the sequencing data can be integrated with structural and functional data, gene expression data, and clinical data. Such a large amount of data poses an urgent need for the development of theoretical algorithms and software in the storage, acquisition, processing, browsing, and visualization of biological information. Since most of our students are biology majors, we will not go into depth about machine learning; we will introduce some common machine learning concepts and algorithms used in biology fields.

In machine learning, there are two different approaches referred to as supervised and unsupervised learning. Supervised learning methods produce models based on known classified data, for instance, a group of genes known to be involved in a certain process. Supervised learning requires training data (the information used to set up the model, that is labelled (meaning the true correct classification of data items is known. Unsupervised methods are applied when we have no gold standard data (labeled data) on which to train a model. We will focus on methods in unsupervised learning in this lab, including cluster analysis and dimensionality reduction (principal component analysis).

Unsupervised learning groups items based on their similarity without knowing why they are related. Basic differential gene expression analysis is an unsupervised procedure—we group genes together based on the difference in their expression in two conditions (control and treatment). More complicated experiments, such as those with multiple treatment groups or time course experiments, are more difficult to interpret. The hope is that we can deduce biological meaning behind the observed differences in a differential expression experiment by grouping similar genes together and looking at them jointly.

Cluster Analysis

In cluster analysis, we group items based on a feature vector – a vector of measurements of properties for each item in the analysis. For genes, the feature vector is the expression level of the gene across the studied conditions. For samples, the feature vector is the relative expression of each observable gene under each condition. These feature vectors correspond to the rows or columns of a data matrix of gene expression levels, where the columns represent experimental conditions and the rows represent genes.

One of the simplest visualizations of the expression data matrix is a heatmap, wherein expression levels are indicated by colors (Fig. 12.1). Generally, you will see heatmaps where clustering has been applied in both the sample and gene dimensions (Fig. 12.1). The method of calculating the distance between feature vectors strongly affects the results. Some common distances include Euclidean distance (Eq. 12.1) and Manhattan (city block distance) distance (Eeq. 12.2), both of which mostly look at magnitude, and are usually used for case-control type experiments. Correlation

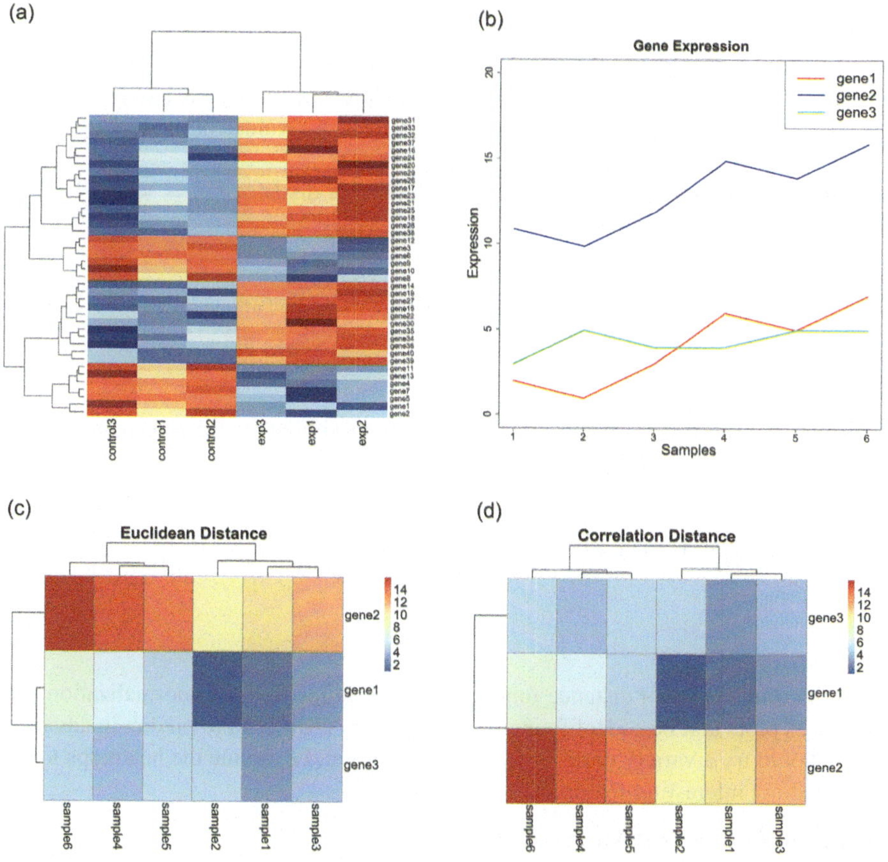

Fig. 12.1 Cluster analysis and the effects of choice of distance function. (**a**) Heatmap of cluster analysis using 40 differentially expressed genes. Rows are genes, columns are samples. Each condition (control and experimental) has three replicates. (**b**) Line graph of three genes expression level. (**c**) Heatmap of cluster analysis using Euclidean distance. (**d**) Heatmap of cluster analysis using correlation distance

distance, such as Pearson correlation (Eq. 12.3), which looks for genes (or samples) whose expression rises and falls in step, is usually used for time course analysis.

For gene expression data, you should use log-transformed expression values in the heatmap because of the very large range observed in gene expression values (about 10,000-fold). It is also common to normalize the values across the rows or columns so that values have a similar range. Gene expression values are usually normalized to have a mean of zero and a standard deviation of one (standard normal deviate).

Euclidean distance: used to cluster genes with similar magnitude of values across the feature vector

$$D(x,y)=\sqrt{\sum_{i=1}^{p}(x_i-y_i)^2} \tag{12.1}$$

where x and y are the items being compared (genes expression values or samples). Values are summed over the feature vector, which has p elements. The Euclidean distance is often dominated by a few values in the feature vector that have the largest differences.

Manhattan distance: works similarly to Euclidean distance but is somewhat less sensitive to large outliers.

$$D(x,y)=\sum_{i=1}^{p}|x_i-y_i| \tag{12.2}$$

Pearson correlation: used to cluster genes with similar expression patterns, but potentially very different magnitudes

$$D(x,y)=1-r(x,y)=1-\frac{\sum_{i=1}^{p}(x_i-\overline{x})(y_i-\overline{y})}{\sqrt{\sum_{i=1}^{p}(x_i-\overline{x})^2\sum_{i=1}^{p}(y_i-\overline{y})^2}} \tag{12.3}$$

The combination of distance function, clustering method, and normalization that produces plots that best identify interesting patterns is difficult to predict in advance. You should try a variety of different combinations and examine the heatmaps to see which show interesting features.

Principal Component Analysis (PCA)

Large complex feature vectors, such as those seen in gene expression analysis, frequently contain many highly correlated elements. In a gene expression experiment these would be sets of genes that respond similarly to the treatment. In a typical differential gene expression analysis, many genes may have relative expression levels that are similar across all experimental conditions. This distorts the pattern seen in clustering because the expression values are not independent, and the clustering result tends to be dominated by the largest groups of correlated values.

PCA tries to solve this problem by transforming the data to a new coordinate system in which the elements of the feature vector are uncorrelated (sometimes this is called independent or orthogonal, the meaning is the same). The new coordinate system (referred to as the principal components) is a weighted linear combination of the elements of the original feature vector, and the new feature vector is generally much smaller. Because of this reduction in the size of the feature vector, PCA is often referred to as a dimensionality reduction method. The mathematics behind PCA involves extracting the eigenvectors and eigenvalues from the data covariance matrix. These methods are taught in linear algebra courses and will not be discussed here.

You can think of PCA as rotating your data onto a new set of axes where the first dimension contains the greatest variance, the second dimension, the second highest variance, and so on. If we are looking at the relationships between six samples, each with a feature vector of 15,000 genes, PCA will produce a set of 6 transformed coordinate axes (Fig. 12.2). Not all of these new axes are equally important—the ones that represent (or explain) the largest amount of the original data variance best capture the relationships in the original data. PCA programs sometimes show only the first two principal components. Look carefully at the sum of the variance explained by the axes; if it is low (<70%) you should look at more than the first two principal components.

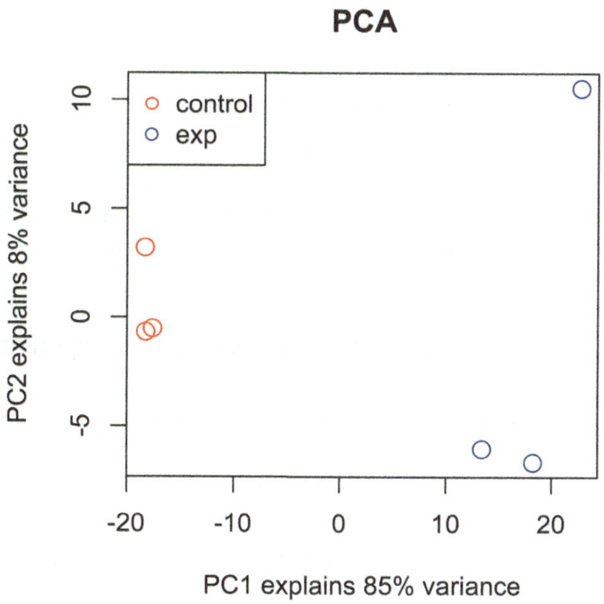

Fig. 12.2 Principal component analysis (PCA) of six example data samples. The red dots represent the three replicates of control group and blue dots represent the three replicates of the treatment group

PCA is usually performed only on the elements of the feature vector that have the highest variance, *i.e.*, those whose values differ the most across the data matrix. The number of most variable elements that are included has a big influence on the analysis. PCA also tends to be dominated by the values that have the biggest magnitude (highest expression values). This is because even random variation in high-expression genes may be larger in magnitude than the measured expression levels of low-expression genes showing systematic effects. For gene expression data, this effect can be somewhat reduced by using log-transformed expression values.

One valuable use of PCA is to verify that the replicates in your treatment groups actually behave similarly. This is done by performing PCA on the sample covariance as shown in Fig. 12.2, where you can see that the control and experimental groups cluster together. You can also see that axis 1 (PC1) clearly separates the control and treatment groups, while there is extensive overlap between the groups along axis 2 (PC2). The most heavily weighted genes that define axis 1 are therefore the ones that best distinguish the treatment groups, while the heavily weighted genes that define axis 2 do not. The total variance explained by the first two principal components is 93%, which indicates a good capture of the variances.

12.3 Before Lab Questions

Using the introductory material and/or online resources, define the following:

1. Supervised learning
2. Unsupervised learning
3. Hierarchical cluster

12.4 Resources and Materials

- R software.
- A raw count matrix: raw.genes.matrix (supplemental file 4).
- A DEG matrix with counts: DEG.csv (supplemental file 5).

12.5 Activity 1: Cluster Analysis and Heatmap

Open the R console and type commands below.

```
# Create an example data matrix with six samples, each with three gene expression values
# the expression level of gene2 is 8 units higher that of gene1 with a small random amount
 (0.0 - 1.0) added
sample1 <- c(2,0,3)
sample2 <- c(1,0,5)
sample3 <- c(3,0,4)
sample4 <- c(6,0,4)
sample5 <- c(5,0,5)
sample6 <- c(7,0,5)
# create a dataframe with six samples
df <- data.frame( sample1, sample2, sample3, sample4, sample5, sample6,
 row.names=c('gene1', 'gene2', 'gene3'))
df['gene2',] <- df['gene1',] + 8 + runif(n=1, min=0, max=1)
# look at what df is
df
# plot the expression of each gene in a different color. First plot gene1, then add lines for gene
 2 and gene3, and the legend
plot(x=1:6, df['gene1',], type="l", col="red",lwd=3, ylim=c(0,20), xlab="Samples",
 ylab="Expression", main="Gene Expression")
lines(x=1:6,df['gene2',],col="blue",lwd=3)
lines(x=1:6,df['gene3',],col="green",lwd=3)
legend("topright", col=c("red","blue","green"), paste("gene",1:3, sep=""),lty=1, lwd=3,
 cex=1.0)
```

```
# plot heatmap with clustering in both dimensions.
# Options:
# scale = "row" or "column" -> normalize the data across rows or column
# clustering_distance_rows_"euclidean", "correlation", "manhattan" -> distance between items
# clustering_distance_columns_"euclidean", "correlation", "manhattan" -> distance between
 items
# clustering_method = "average", "ward.D2", "complete" -> method for forming clusters
library(pheatmap)
pheatmap(as.matrix(df), scale="none",
 clustering_distance_rows="euclidean",clustering_distance_columns="euclidean",clustering_
 method="average",cellwidth=80, cellheight=120, main="Euclidean Distance")
pheatmap(as.matrix(df), scale="none",
clustering_distance_rows="correlation",clustering_distance_columns="correlation",clustering
_method="ward.D2",cellwidth=80, cellheight=120, main="Correlation Distance")
```

12.6 Activity 2: Principal Component Analysis Using all Genes

```
# prcomp-based on singular value decomposition
# type use ?prcomp for more information
# read data
data1 <-read.delim("/PathToData/raw.genes.matrix", stringsAsFactors=F, row.names=1)
head(data1)
dim(data1)
# transpose data so that rows are samples, columns are features/genes. This will produce a
 PCA plot of samples vs samples. Three different versions of the data are created:
# dataT - original complete data
# dataT2 - only the genes where the sum of the expression across all samples is > 0.1
# dataT3 - log-transformed version of dataT2
dataT <- t(data1)
dataT2 <- dataT[, colSums(dataT != 0) > 0.1]
dataT3 <- log10(dataT2+1)
# in the following section do the analysis with each of the three datasets dataT, dataT2 and
 dataT3
pca <- prcomp(dataT3, scale.=TRUE)
summary(pca)
# scree plots are used to examine the amount of the variance represent by each principal
 component (transformed axis)
barplot(pc$sdev^2/sum(pc$sdev^2), xlab="principal component", ylab="% of variance")
#cumulative variance
barplot(cumsum(pc$sdev^2/sum(pc$sdev^2)), xlab="principal component",
 ylab="cumulative % ofvariance")
```

```
# PCA plot
myColor <- rep(NA,nrow(dataT3))
myColor[1:3] <- "red"
myColor[4:6] <- "blue"
plot(pc$x[,1:2],col=myColor,pch=16,cex=2,main="PCA",xlab="PC1", ylab="PC2")
legend("topleft",pch=16,col=c("red","blue"), c("control","exp"))
```

12.7 Activity 3: Principal Component Analysis Using DEG

```
# read data
data2<- read.delim("/PathToData/DEG.csv" ,stringsAsFactors=F,row.names=1,sep=",")
head(data2)
dim(data2)
dataT <- t(data2[,8:13])
dataT2 <-  dataT[, colSums( dataT != 0) > 0.1]
dataT3 <- log10(dataT2+1)
pc2 <- prcomp(dataT3,scale.=TRUE  )
summary(pc2)
#scree plot
barplot(pc2$sdev^2/sum(pc2$sdev^2),xlab="principal component",ylab="%ofvariance")
#cumulative variance
barplot(cumsum(pc2$sdev^2/sum(pc2$sdev^2)),xlab="principal component",
ylab="cumulative % ofvariance")
# score plot
myColor  <- rep(NA ,nrow(dataT3))
myColor [1:3] <- "red"
myColor [4:6] <- "blue"
plot(pc2$x[,1:2],col=myColor, pch=16,cex=2,main="PCA" ,xlab="PC1",
ylab="PC2" ,cex.lab=1)
legend("topleft",pch=16,col=c("red","blue"),legend=c("control","exp"), cex=0.9)
```

We may also save figures in either JPEG format or TIFF format.

```
# save figure in jpeg format
jpeg(filename="/PathToFig/Fig_PCA.jpeg" , width=5400, height=5400, res=900)
par(mar = c( 5,5,4,2) + 0.2)
plot(pc2$x[,1:2],col=myColor, pch=16,cex=2,main="PCA" ,xlab="PC1",
ylab="PC2" ,cex.lab=1)
legend("topleft",pch=16,col=c("red","blue"), legend=c("control","exp"), cex=0.9)
dev.off()
```

12.8 Analysis/Notebook

- Based on the examples and instructions above, please analyze the DEG.csv (supplemental file 5) using clustered heatmaps and PCA methods.
- When you use all data to do PCA, are the first two principal components sufficient, why or why not?
- When you only use DEG data to do PCA, is this a better result? Why or why not?

Further Reading

1. StatQuest: principal component analysis (PCA), step-by-step. https://www.bilibili.com/video/av69646206
2. A step-by-step explanation of principal component analysis (PCA). https://builtin.com/data-science/step-step-explanation-principal-component-analysis
3. Introduction to clustering in machine learning: types, algorithms, and applications. https://hackernoon.com/introduction-to-clustering-in-machine-learning-types-algorithms-and-applications

The manufacturer's authorised representative in the EU is Springer Nature Customer Service Centre GmbH, Europaplatz 3, 69115 Heidelberg, Germany. If you have any concerns regarding our products, please contact ProductSafety@springernature.com

Printed and bound by CPI Group (UK) Ltd, Croydon, CR0 4YY

01/10/2026

02225880-0014